Brownfield Sites III

Prevention, Assessment, Rehabilitation and Development of Brownfield Sites

WIT*PRESS*

WIT Press publishes leading books in Science and Technology.
Visit our website for the current list of titles.
www.witpress.com

WIT*eLibrary*

Home of the Transactions of the Wessex Institute.
Papers presented at Brownfields III are archived in the WIT eLibrary in volume 94 of WIT Transactions on Ecology and the Environment (ISSN 1743-3541).
The WIT eLibrary provides the international scientific community with immediate and permanent access to individual papers presented at WIT conferences.
Visit the WIT eLibrary at www.witpress.com.

THIRD INTERNATIONAL CONFERENCE ON PREVENTION, ASSESSMENT, REHABILITATION AND DEVELOPMENT OF BROWNFIELD SITES

Brownfields III

Conference Chairmen

C. A. Brebbia
Wessex Institute of Technology, UK

Ü. Mander
University of Tartu, Estonia

International Scientific Advisory Committee

Organised by
Wessex Institute of Technology, UK
Tartu University, Estonia

Sponsored by
The International Journal of Sustainable Planning and Development

WIT Transactions on Ecology and the Environment

Transactions Editor

Carlos Brebbia
Wessex Institute of Technology
Ashurst Lodge, Ashurst
Southampton SO40 7AA, UK
Email: carlos@wessex.ac.uk

Editorial Board

Brownfield Sites III

Prevention, Assessment, Rehabilitation and Development of Brownfield Sites

Editors

C. A. Brebbia
Wessex Institute of Technology, UK

Ü. Mander
University of Tartu, Estonia

WIT*PRESS* **Southampton, Boston**

C. A. Brebbia
Wessex Institute of Technology, UK

Ü. Mander
University of Tartu, Estonia

Published by

WIT Press
Ashurst Lodge, Ashurst, Southampton, SO40 7AA, UK
Tel: 44 (0) 238 029 3223; Fax: 44 (0) 238 029 2853
E-Mail: witpress@witpress.com
http://www.witpress.com

For USA, Canada and Mexico

Computational Mechanics Inc
25 Bridge Street, Billerica, MA 01821, USA
Tel: 978 667 5841; Fax: 978 667 7582
E-Mail: infousa@witpress.com
http://www.witpress.com

British Library Cataloguing-in-Publication Data

A Catalogue record for this book is available
from the British Library

ISBN: 1-84564-041-1
ISSN: 1746-448X (print)
ISSN: 1743-3541 (online)

The texts of the papers in this volume were set individually by the authors or under their supervision. Only minor corrections to the text may have been carried out by the publisher.

Printed in Great Britain by Athenaeum Press Ltd., Gateshead.

Preface

This book contains most of the papers presented at the 3rd International Conference on Brownfield sites: Assessment, Rehabilitation and Development, held in Tallinn, Estonia, in 2006, organised by the University of Tartu and the Wessex Institute of Technology. Previous meetings in the series were held in Cádiz, Spain (2002) and Siena, Italy (2004).

Given the economic and social benefits of brownfields redevelopment, there is a need for guidance on a process of ensuring the acceptability and therefore viability of such redevelopment. The preparation of the guidance requires further research as well as the sharing of information, lessons and experience among experts in this field.

This volume brings together international information, experience and research in order to give the reader a greater knowledge and ability to help revitalize their communities. It discusses some of the problems facing the public and private sectors and the engineering and scientific communities. It studies long-term plans for the productive re-use of lands and properties that have been abandoned or lie idle, and looks at some of the financial, regulatory and community interests which complicate the whole process.

The papers published in the book are grouped into the following sections: Development issues; Environmental assessment; Risk assessment and management; Cleanup methodologies; Case studies; Community and public involvement; Lessons from the field; Modelling and assessment, and Monitoring of contaminated sites.

The Editors are grateful to all authors for their contributions and to the members of the International Scientific Advisory Committee for their help in selecting the papers included in this Volume.

The Editors
Tallinn, 2006

Contents

Section 4: Cleanup methodologies

Section 5: Case studies

Section 6: Community and public involvement

Section 7: Lessons from the field

Section 8: Modelling and assessment

Section 9: Monitoring of contaminated sites

Section 1
Development issues

Re-using urban industrial areas:
the case of 's-Hertogenbosch, the Netherlands

W. van der Toorn Vrijthoff
Delft University of Technology, Faculty of Architecture, Department of Real Estate and Project Management, Delft, The Netherlands

Abstract

In the 20th century, and especially after World War II, European cities began to grow very rapidly. Historic inner-city areas currently occupy less than 5% of the total urban area of the old cities. This implies that the industrial sites dating from the beginning of the 20th century are no longer situated on urban peripheries, but are to be found in the central areas of cities. A consequence of this growth was that the old city centers became far too small to meet the needs of these enlarged cities, so there was also a growing need to expand the centers.

To meet this need, many of the old urban industrial sites in European cities have been transformed into functional extensions of the old inner cities. In some cases, the old buildings and infrastructure have been included in the new plans to retain the historic character of the area; in other cases re-development has brought about a total transformation, making it hard to recognize anything of the historic appearance in the new one.

This paper focuses on 's-Hertogenbosch, an old city in the southern part of the Netherlands that gained its charter just before 1200. Close to the city centre there was an old industrial area being transformed into "a monument for the future". It will become a new part of the city centre, providing 1,400 houses and 180,000 m^2 of office space. In addition, 30,000 m^2 of space will be reserved for hotel and catering businesses, retail outlets and culture. This paper tells the story of how this industrial area is being transformed; the vision behind it, the plans, and how they will be carried out. Special attention will be given to how these plans relate to, or ignore, features of the historic site.

Keywords: urban transformation, city centre, industrial areas, public–private partnership.

WIT Transactions on Ecology and the Environment, Vol 94, © 2006 WIT Press
www.witpress.com, ISSN 1743-3541 (on-line)
doi:10.2495/BF060011

1 Urban growth and transformation

In Europe, and also in the Netherlands, in the period following World War II, building work was primarily directed towards repairing war damage. The next priority, after that, was to create living space for the steadily growing population. The steadily decreasing average number of people per household, combined with a favourable economic climate, resulted in a constantly growing demand for quantitative space. In response to this demand, building production was mainly oriented towards extending the urban area. In fact, 75% of current urban real estate was built in the last 60 years by transforming rural into urban areas.

1.1 Increase in scale and concentration

The spatial functional concepts on which urban-area extension was based in the decades following World War II were greatly influenced, from 1929 onwards, by ideas submitted by the Congress International d'Architecture Moderne (CIAM) [International Congress on Modern Architecture] (Ministry of Housing, Spatial Planning and the Environment, 2004). At least, that is the generally accepted explanation for the realization of large post-war housing areas where the spatial functions are separated. This monofunctionality, in combination with uniformity, is now recognized as one of the main reasons why these areas are so unattractive. The question is whether it could have been done differently. A variant of this question arises in connection with the restructuring assignments for those areas. Is it possible to counterbalance the monofunctionality and uniformity by restructuring these areas to transform them into multifunctional areas with a strongly differentiated form? That would be difficult, if not impossible. The monofunctionality is not a characteristic that, in hindsight, we can ascribe entirely to the CIAM. On the one hand, monofunctionality is largely a consequence of a very strong growth in the use of living space and, on the other, on developments aimed at concentration and scale increase in other sectors. These trends manifested themselves strongly, especially during the second half of the 20th century, and are still present. They have led to the segregation of urban functions, which implies that the traditional mix of functions has become restricted to a limited number of urban areas. Good examples of this are the historic inner city and the urban centre, both of which have long been viewed as areas characterized by a broad mix of functions. These are areas where the functional profile has changed drastically in recent decades, because it was accompanied by the need for expansion.

1.2 The changing profile of the city center

The post-World War II period was a time of renewal throughout Europe. For a decade, it was supported by sustained economic growth. This was coupled by a steady growth in car ownership. The historic inner city was an area in which all urban facilities were concentrated, but accessibility to that area was becoming more and more of a problem. The first reaction to the exodus of commercial activity was to concentrate on modifying the spatial structures, thereby adapting the historic centers to the requirements of the time. The changes were particularly rigorous in the United States. Reacting to that development, Jane Jacobs [1] noted that the unique qualities of old city centers were rapidly being effaced. Though rather more slowly, and not quite as comprehensively, the same thing could be seen happening in Europe.

It soon became clear that the suburbanization of economic life could not be stopped by metamorphosing urban spatial structures. A division of economic interests took place, which caused the historic centre to change from being a place of production to one of consumption. A repetition of this process is imminent today.

For years, the geographical centre of gravity of consumer expenditures had been located in the city centre. Because growth in consumer expenditure is coupled with a substantial increase in spatial requirements, this is now changing. A self-evident, but ill-advised, reaction to this trend is to instigate a far-reaching re-organization of the city centre so as to create more space. By so doing, there is a high probability that the unique spatial quality of the old centre will be destroyed, and that it will become a poor alternative to locations that are both more spacious and more accessible. Instead of taking drastic measures to transform old centers, extending them is a better alternative. Such extensions must then meet the spatial requirements of businesses and individuals who prefer to be located in urban centers. In many cities, such locations are present in the form of industrial sites that date from just before or just after World War II. These areas were built on the edges of cities, often near a railway, but nowadays, because of urban expansion these same areas lie in the heart of urban centers.

1.3 Area market misfit and transformation

Transforming an existing urban area interferes with the societal functioning of that area. This implies that there are interests at stake that extend beyond those of the local parties who are directly involved. It also implies that, in many cases, it is local politics that set the conditions for the decision-making process on a local level. Shortly after the war, the political aims behind the extension of urban areas were not only unambiguous but also similar. Opinions nowadays are often divided and the aims are incompatible. Attempts to reach a consensus about a spatial/functional intervention on a local level are often very time consuming. There is always the danger that, with changes in the balance of power within local politics, directional changes will take place that will seriously frustrate the process of furthering restructuring. A large number of parties can be identified at local level, ranging from owners to users and from public to private parties, all of which have different interests, aims and means. The extent to which all these parties benefit from local transformation differs widely. This provokes discussion, frequent lengthy negotiating procedures and active resistance from some of the parties involved.

Neither the current functioning of an area, nor the immediate environment should be inconvenienced during a process of spatial intervention. It is a matter of 'repairing a running engine' and that is an extremely difficult task. There is often a series of smaller, connected problems, between which interaction occurs.

These smaller problems cannot be solved either separately or one after the other, because the solution to one partial problem affects a series of other partial problems. All these smaller problems therefore have to be approached as a totality of linked parts, because integrated partial problems require integral solutions.

Comprehensive area development is seen as the answer to comprehensive area–market misfit

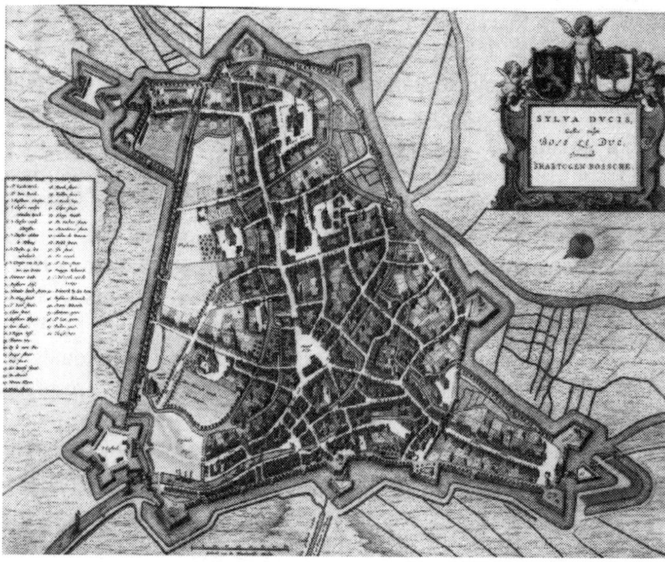

Figure 1: A historical map of the city of ’s-Hertogenbosch. Source: Blaeu’s *Toonneel der Steden* from 1652.

2 The Case of ’s-Hertogenbosch

2.1 The historical background

’s-Hertogenbosch is an average-sized Dutch city, dating from the 13th century. It currently has a population of 132,000. The medieval structure of the historical part of the city, with its ramparts and pattern of narrow streets, has been largely preserved. The Binnendieze, a historic watercourse that runs partly underground through the city, is also still there, thus confirming the authenticity of that part of the historic centre. The old city is situated on a sandbar, surrounded by marshy land. Because of this, urban expansion was very problematic in the past. In addition, ’s-Hertogenbosch was a fortified city with impressive, sturdy ramparts. The city plan of 1652, showed in figure 1, is still very recognizable in the present city plan. In the 19th century, the city was forced to seek solutions for the growing need for space by opting for a high concentration of buildings within the city walls. It was only in 1874, when a law was passed stating that the fortifications of Dutch towns and cities were no longer functional, that creative solutions were sought for making the surrounding marshy land suitable for building. The railway and the station were built on the periphery of the historic inner city, an area in which 590 national and municipal monuments are situated. These buildings strongly determine its authenticity, historical character, and therefore the attractiveness of this area. For this reason, the area has a protected status, so that it is difficult to realize any spatial and functional interventions. "Any urban planning, in almost any European city, cannot now occur other than within the constraining context of the conserved urban landscape" [2]. The inner city also plays an important role in the economic functioning of the city, because more than a third of the employment is situated here. 's-Hertogenbosch has maintained and strengthened its past role as an urban centre, however, not without the growing need to extend it.

2.2 The need to extend the city centre

Because of the economic growth of 's-Hertogenbosch, an ever greater pressure was put on the inner city to accommodate offices, shops and housing. Ever since the 1980s, the municipality had been searching for possibilities to extend the inner city.

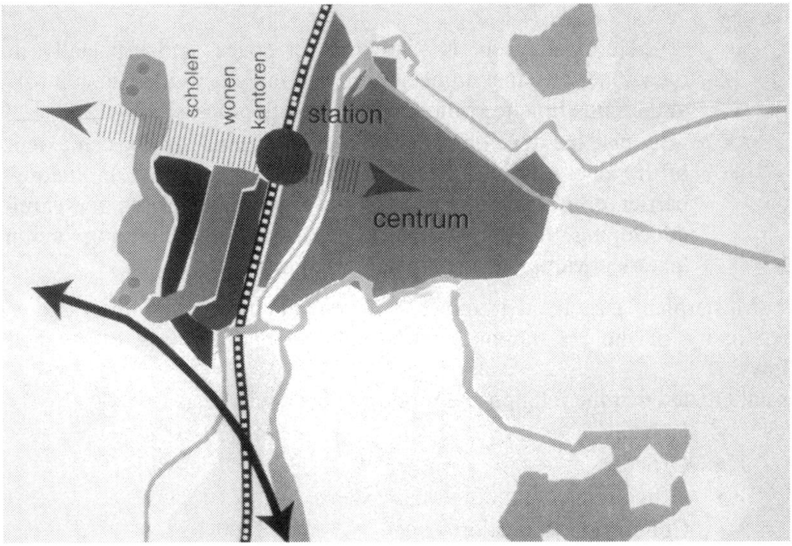

Figure 2: Scheme of the city and the extension to the west.

At the end of the 1980s, the decision was taken to extend the inner city in a westerly direction. The challenging title given to this risky undertaking was the 'leap over the railway'. On the western side of the railway was an industrial site that had been developed around the war years. Many of the businesses that had originally occupied this site had moved, in the meantime, to locations that offered more space, were more accessible and where carrying out the business was less of a nuisance for those in the immediate surroundings. Existing buildings started to become empty in the 1980s. This offered perspectives for large-scale redevelopment. Figure 2 is giving a schematic view of the extension.

The main specification for redeveloping the area around the railway was to 'lift' the inner city over the railway to create a place in the western section of that zone for city-centre functions that could not be accommodated in the historic inner-city due to lack of space. Such was the location, that it offered possibilities to create a new, high-quality urban environment, with new, high-density architecture, and modern solutions. An additional specification was to create space for the new 'spatial claimants' on the market – large-scale organizations in education, the service industries and banking – that had resulted from mergers, the functions of which, as far as scale level was concerned, not longer fitted into the finely meshed parceling out of the historic inner-city.

2.3 Functional program and flexibility

A number of characterizations of the area were included in the points of departure, in the first part of the Masterplan. These are characterizations that are directional for the market position aimed at within the urban context. Of particular importance in this respect is the functional relation with the historic inner-city.

- " 's-Hertogenbosch, is searching for space, and especially in its busy inner city that attracts more and more activities, to control, but at the same time to stimulate, this growth process" and
- "Because the development of the railway zone involves an extension of the inner city by functionally redeploying and by eliminating the barrier effect of the railway, the opportunity arises for carefully developing the characteristic inner-city by utilizing cultural–historical values" (Masterplan, 1996) [3].

The Masterplan largely determines the spatial quality of the area. The functionality of the programme that has to be realized was established as follows:
It is anticipated that the relations between functions will be as follows:

- Housing 30-50%
- Offices 20-40%
- Office-related buildings/showrooms 10-25%
- Culture and Education 0-10%
- Commercial functions 2-5%.

These are margins that are large enough to accommodate the radically different functional programmes of variants in the plans. Market developments during the formation of plans have been determining for optimizing the functions within the margins indicated. That has resulted in a reduction in the office share compared with the initial plans and an enlargement in the share of the housing function.

Flexibility has thus made it possible to move with the market, contributing towards a financially positive end-result.

	Station Square*	Paleiskwartier 1st phase*	2nd phase	Total
Surface (hectares)	8	10	18	36
Offices (m²)	10.300	91.000	110.600	211.900
Other commercial functions (m²)	6.800	6.300	32.700	45.800
Dwellings (Number of)	16	307	1.200	1.523

Figure 3: The functional programme.

Figure 4: The redevelopment area.

During the plan formation, ambitions with respect to concentration and urban quality were adjusted to ever higher levels. That has brought an area into existence that many people view as the modern equivalent of the historic inner-city of 's-Hertogenbosch. Figure 3 shows the realized functional programme and in figure 4 the redevelopment plan is shown.

2.4 Public private partnership

The choice made for the Palace District (Paleiskwartier) was a far-reaching public–private partnership. The municipality was active in developing the plan, together with three commercial parties. The utilization of both land and buildings was coordinated under the Paleiskwartier BV, the limited company set up to carry out the Palace District project, in which each of the four parties had an equal, 25%, share of the ownership.

Figure 5: The area in transformation, centre, the water basin with the underground parking facility.

The early acquisition of land stimulated the process and gave an advantage over other parties. This prevented encroachment on ownership, and also speculative purchasing by other private parties which could have forced up the prices. Had no action been taken, then other (unwanted) parties would have been able to acquire claims on land in the area. Early acquisition was thus essential. The fragmented ownership of land was breached, which gave rise to an important condition for integral area development and also to the conditions within which urban design aspirations and ambitions for the planning area could be achieved. The redevelopment of the area was set up in phases because, at the time when the plan started to be carried out, part of the area was still in use. This is why the decision was made to base the new urban architectural structure on the one that already existed. Although the area as a whole was totally redeveloped, it was nevertheless decided to build an underground parking garage, under the elongated water basin, in the central part of the planning area. Figure 5 is showing the area in transformation with the water basin in the center of the area. This provided an underground solution for parking pressure in the area, and had a positive effect on the quality of public space.

Concealed under the water feature, there is now a public parking garage, with enough space for 1070 vehicles. In this way, public space in this intensively built-up area has been qualitatively designed, so that only a limited number of parking places (approximately 200) are situated on ground level [3].

3 Obsolete industrial sites in the Netherlands

It has been estimated that on 1 January 2003 there was a gross area of 93,800 hectares of industrial land in the Netherlands, divided between 3600 business sites (IBIS, Integral Bouw Informatie Systeem) [4]. Of this, 24,400 hectares were open ground used as, or earmarked for, public use, infrastructure, green facilities, etc. The net remaining area of 69,500 hectares is built-up land. The same research also revealed that, of the 93,800 hectares of industrial land, 30% is obsolete and eligible for redevelopment. This is approximately 30,000 hectares, which represents a building location 800 times as large as the location in 's-Hertogenbosch, described earlier in this paper. This location is large enough to accommodate the expanding needs of urban functions, including housing, for very many years to come.

References

[1] Jacobs, J., 1961, The death and life of great American cities, New York
[2] Ashworth, G.J., 2002, Modern Europe, Arnold Publishers, London
[3] Bruil, I. et al., 2004, Integrale gebiedsontwikkeling [Integral area development], Amsterdam, SUN publishers.
[4] IBIS, 2003, Werklocaties 2003:Hoofdrapportage, Den Haag

Recycling brownfields? The influence of on-site contamination on home buyers

K. Winson-Geideman
Department of Landscape Architecture and Urban Planning, Texas A&M University, USA

Abstract

This research uses contingent valuation (CV) to assess home buyer perceptions regarding housing built on formerly contaminated sites. Respondents are questioned about the underlying stigma associated with past contamination and willingness to purchase a home built on a brownfield. Discounts and market loss in Chicago, Illinois, a metropolitan area characterized by a large number of brownfields and an active housing market, are identified and analyzed.
Keywords: brownfields, contingent valuation, stigma.

1 Introduction

Over the past 50 years, tens of thousands of the factories, warehouses, rail yards, and other facilities that stimulated industrial growth in many parts of the world have been abandoned. Buildings became obsolete as innovation required the use of new technologies and companies moved to new locations. Many of these facilities were in use before there was significant regulation of hazardous substances, and quite often are now impacted by residual contamination though the type and extent is unknown. Many times the party responsible for the contamination is no longer in existence.

The high number of these sites, known as brownfields, and the associated threats due to contamination often result in negative impacts on property value and greater risk for property investment. While a substantial amount of research addresses value and liquidity issues when these sites are returned to industrial use or reclaimed for commercial purposes, in general, there are few studies that address the demand for residential use, or how individual homeowners might integrate knowledge of contamination into location decisions.

WIT Transactions on Ecology and the Environment, Vol 94, © 2006 WIT Press
www.witpress.com, ISSN 1743-3541 (on-line)
doi:10.2495/BF060021

Prior research conducted by the same author shows evidence that current disclosure methods, typically deed restrictions, do not provide sufficient notice to individual consumers about contamination, use restrictions, or maintenance procedures associated with remediated property [1]. Individuals typically have little knowledge of the potential threats to health and financial investments while businesses and public institutions employ a bevy of legal and scientific professionals to determine the risk of owning and redeveloping brownfields. If maintenance procedures are then neglected, containment measures are rendered ineffective making human exposure to potentially toxic chemicals much more likely.

The purpose of this study is to evaluate the impact of contamination on the location decision of potential home buyers, specifically where the home is built on a site that has completed a state-sponsored, brownfield cleanup program. Contingent valuation is used to estimate the perceived risk associated with such sites.

2 Literature

Shifting economic conditions and other factors, have forced many states and municipalities to rethink their approach to brownfields. Many times, advances in transportation and policy changes have made the sites no longer viable for industrial development. Some find new life as commercial/retail or industrial developments, but others are poorly situated for such use due to size, location and other factors. Past U.S. brownfield policy has been directed to economic development and has thus ignored the potential of converting sites into housing, even though in some situations this is the highest and best use.

More recent trends, however, show that state Voluntary Clean-up Programs (VCPs) are beginning to actively encourage redevelopment for residential use. Since many of these sites are located in easily accessible, urban areas within close proximity to entertainment, educational opportunities, and infrastructure, it is a reasonable and logical use. In a report released in 2000, 5,200 units of housing had been built on brownfields in California, and 2,855 and 1,400 were constructed in Colorado and Michigan, respectively [2].

Given that, a certain degree of residual risk or "stigma" is likely to remain on sites that are subject to past contamination issues or containment measures. Typically the stigma surrounding a contaminated site is greater prior to cleanup, although residual stigma may remain, manifest in failed or discounted transactions. Mundy classifies the consequences of stigma into two areas: direct and indirect [3]. While increasing incidence of cancer or decreased occupancy in an apartment building exemplifies direct effects, the exodus of residents from an effected area, such as the Love Canal in New York is an example of an indirect effect.

Studies show that knowledge of contamination and residual stigma negatively impact property markets. In a study of the impacts of PCB contamination on property values in Anniston, Alabama, Simons found 95% of potential bidders were unwilling to bid on contaminated property [4]. Simons, Winson, and

Mikelbank note that an oil spill on the Patuxent River in Maryland increased market time for properties by 40% [5]. McCluskey and Rausser [6] found that increased information from media coverage affected property sales. Additionally, the reluctance of lenders to provide funding for remediated sites may impact the number of bidders [7].

This study uses contingent valuation to assess the discount and number of failed transactions associated with residual stigma caused by contamination. It is anticipated that if home buyers are informed of prior contamination before a purchase contract is executed, they will either discount the transaction based on their perception of risk or they will withdraw from the market altogether.

3 Methodology

Within the past ten years, contingent valuation has become an important tool in the valuation of environmental damages for market goods. Originally used to value non-market goods, particularly natural resources, it involves the use of formal surveys of market participants to determine willingness to pay or accept compensation for damages. When used to evaluate the impact on market goods, the same survey methodology is used, but the *discount* associated with damages and the *willingness to bid* of the respondents are the best methods of measurement.

Simons [4] recognizes that CV can provide guidance when valuing contaminated property. The author recommends the use of two moderately rigorous techniques, rather than a single strong one when data permits. Chalmers and Roehr [8], Syms [9] and McLean and Mundy [10] all recommend the use of CV, though the former suggest it be used in concert with the results of a regression analysis (within a hedonic pricing framework). The results of the regression can then be used to verify the findings from the contingent valuation analysis.

3.1 Survey instrument

The survey instrument and methodology focuses on gauging a respondent's willingness to bid on a remediated site and any discount that may be associated with that bid. It adopts the survey and CV methodology used by Simons [4] in a case study of PCB contamination in Anniston, Alabama [11].

The first section of the survey is designed to acclimate the respondent to the survey's subject and to the scale used to measure responses (-3 to +3 with 0 as neutral). A question is included that establishes the value of the respondent's current home which is used as a baseline response to evaluate any discount associated with contamination.

The second section contains questions that place the respondent's current home in a broader context with three scenarios that may also be interpreted as environmental disamenities. All describe the respondent's current home, yet place it within direct proximity to different types of environmental disamenities: one block from a business park; adjacent to a recently remodeled, operating attractive gasoline service station with leaking underground storage tanks (LUST); and an on-site deed restriction limiting ground water use and requiring a building to stay in place as a permanent cap over contamination. Respondents are then asked to value the property given the changes indicated in the scenario.

The last scenario serves as the primary source for data on the stated preferences of the respondents regarding purchasing a home on remediated property. The scenario is replicated from the deed of a property in Chicago that was redeveloped residentially and indicates the home was "part of a state mandated environmental clean-up program. To successfully complete the program, several feet of soil had to be removed from the site and replaced with clean soil. Public record now shows that it has been cleaned to residential standards and is suitable for humans to live on, although there is a restriction indicating that the building must remain over the formerly contaminated portion as a permanent cap. This restriction is a part of the deed and will be passed on in perpetuity."

The questionnaire was administered by telephone and 74 responses were gathered. Three surveyors collected the information over several months in 2002. The surveyors were instructed not to supply information or opinions to the respondents beyond what was included in actual instrument. One survey was omitted from the final analysis because the administrator failed to collect proper data.

3.2 Measurement techniques

Two factors are important when evaluating the descriptive results of contingent valuation. First, the portion of respondents willing to bid on a scenario is reflective of market demand. Willingness to bid is indicated by a response to the relative contamination scenario that is greater than $0. If the respondent answers $0, it is assumed that he or she is *unwilling* to bid indicating a loss in a market participant. In the event the respondent states a value greater than $0, he or she is considered *willing* to bid, thus showing a willingness to participate in the market for remediated land.

Second, the ratio of maximum bid to the baseline case value reflects the percentage discount of those that did bid. This number is calculated as one minus the percentage discount in price. For example, if the respondent's baseline price is $100,000 and the maximum they are willing to bid is $40,000, a 60% discount is reflected [4]. This creates an assessment of consumer risk perception by assigning a dollar value and hence a percent change in the price of property due to contamination.

4 Results

The bidders from each of the scenarios were divided into those that were willing to bid (the average bidder), those who bid in the top half of all bidders (top half) and those that bid in the top quarter of all bidders (top quarter). To determine which bidders comprise the top half, the bidders were categorized based on the percentage discount of their bid. The total number of bidders was divided in half based on the bid discount. The portion with the lowest bid discount became the top half. Top half bidders were divided the same way to produce the top quarter of all bidders. This is done to control for those bids that are so small that a reasonable seller would be unwilling to accept them.

In the business park scenario, a majority of the respondents indicated a willingness to bid. The respondents appeared to consider that scenario the least detrimental, with an overall bid percentage of 70% (51 bidders) with an error rate of +/- 7.59%. The average discount of all bidders was 18%.

Interestingly, the top quarter of the bidders were willing to pay a 5% premium for property within one block of the business park. This may indicate that some buyers are willing to pay a premium to live close to potential employment, but not directly adjacent to it. Yet these numbers are somewhat suspect, given that three participants responded with particularly high premiums. When removed from the calculations, the top quarter shows a 1% premium. The responses to the business park scenario are listed in Table 1.

Table 1: Descriptive measurements of business park scenario.

	Number of Responses	Number of bidders	Bid Percent	Average Discount
Average	73	51	70%	-18%
Top Half	37	26	35%	-1%
Top Quarter	19	13	17%	5%

Of those that bid, the LUST scenario was widely considered to be the most debilitating with only 11 of the 73 (15%) respondents choosing to bid on property adjacent to a site with leaking underground storage tanks. The average discount on this scenario was 52% with an error rate of +/-27.42%. The top half of the respondents discounted the property value by 31% and the top quarter of the bidders discounted it by 14%. There were no premium bids for this scenario and every bidder indicated a discounted value. These results are consistent with similar research conducted by Simons and Winson-Geideman [12]. The responses are summarized in Table 2.

Table 2: Descriptive measurements of LUST scenario.

	Number of Responses	Number of bidders	Bid Percent	Average Discount
Average	73	11	15%	-52%
Top Half	37	6	8%	-31%
Top Quarter	19	3	4%	-14%

The final scenario reflects the impact of deed restriction associated with prior contamination on bidding and value. Results show a bid percentage of 38% with an error rate of ±14.64% at a 95% level of confidence, indicating that subsequent responses to the same survey question will be within ±14.64% of what is indicated. The responses fell into the mid-range of perceived disamenities. The average bidder indicated a discount of 39%. The top half and top quarter bid discounts of 15% and 10%, respectively. The results for the deed restriction scenario are in Table 3.

Table 3: Descriptive measurements of deed restriction scenario.

	Number of Responses	Number of bidders	Bid Percent	Average Discount
Average	73	28	38%	-39%
Top Half	37	14	19%	-15%
Top Quarter	19	7	10%	-10%

5 Conclusions

The results of the contingent valuation study show three different discount levels derived from the data that was collected. The average bidder shows a discount of 39%, followed by the top half of bidders with a 15% discount and the top quarter with a 10% discount. The bids are calculated in this manner to control for "bottom feeders" or bidders with such extreme discounts that it is unlikely that a rational seller would accept such a bid. For instance, the overall average includes "bottom feeding" discounts as large as 95%.

The level of discount that would most likely be reflected in the actual market falls between the top half and the top quarter discounts of 15% and 10%, respectively. This level was chosen for three reasons. First, it most closely reflects market studies using multiple regression that quantify the residual effects of contamination on nearby property values. Simons found that uncontaminated, easement-holding properties within two miles of a petroleum pipeline rupture experienced a loss in value of 4–5% [13]. Nelson, Genereux and Genereux found a reduction in the sale price of homes near a landfill of 6–12% depending on distance [14]. While these studies are focused on proximity to contamination, it is likely that homes built directly on a contaminated site will experience a slightly higher discount.

Second, the site described in the scenario is low-risk. Careful review shows that the site had been cleaned to residential standards and the dirty soil was removed. Further, any residual contamination is encapsulated by the building minimizing risk of exposure.

Finally, the site had completed a state-mandated environmental cleanup program and the remaining problem is the residual stigma associated with the history of contamination. This leads to the assumption that similar sites have market potential, and property values remain largely intact although some discount can be expected.

The number of bidders affected by the existence of prior contamination shows a decrease in market demand of 62%. Just over one-third (38%) of the respondents were willing to bid on the deed restriction scenario. Although the market loss is less than in Greenberg and colleagues' study (14%), the results emphasizes the pervasiveness and consistency of the stigma associated with brownfields [15]. Given that, it is likely that buyers will become more willing to purchase brownfield housing as they become more educated about the risks and how they are best minimized.

References

[1] Winson-Geideman, K. Environmental case studies: ensuring suitable comparables." *The Appraisal Journal*, 73(3), pp. 53-61, 2005.

[2] Bartsch, C. and Dorfman, B., Brownfields and housing: how are state VCPs encouraging residential development? Northeast-Midwest Institute, 2000.

[3] Mundy B., Valuing Brownfields (Chapter 6). *Brownfields: a comprehensive guide to redeveloping contaminated property*, eds. Davis, T. and K. Margolis, American Bar Association: Chicago, 1997.

[4] Simons, R., Estimating proximate property damage from PCB contamination in a rural market: a multiple techniques approach. *The Appraisal Journal,* 70(4), pp: 388-400, 2002.

[5] Simons, R., Winson, K. and Mikelbank, B., The effects of an oil pipeline rupture on single-family house prices. *The Appraisal Journal*, 69(4), pp: 410-418, 2001.

[6] McCluskey, J. and Rausser, G., Estimation of perceived risk and its effect on property values. *Land Economics*, 77(1), pp: 42-55, 2001.

[7] Jackson, T. Environmental risk perceptions of lenders and investors in commercial and industrial real estate. Paper presented at The Sixteenth American Real Estate Society Conference, Santa Barbara, California, March 28 to April 1, 2000.

[8] Chalmers, J. and Roehr, S., Issues in the valuation of contaminated property. *The Appraisal Journal*, 61(1), pp. 28-41, 1993.

[9] Syms, P. *The redevelopment of contaminated land for housing use.* Joseph Rowntree Foundation: London, 1997.

[10] McClean, D. G. and Mundy, B., Adding contingent valuation and conjoint analysis to the required body of knowledge for the estimation of environmental damages to real property. *Journal of Real Estate Practice and Education,* 1(1), pp. 1-12, 1998.

[11] The survey is based on an instrument developed by Barton Smith, PhD, University of Houston Department of Economics and adapted by Robert

A. Simons, PhD, Cleveland State University Department of Urban Affairs.

[12] Simons, R. A. & Winson-Geideman, K., Determining market perceptions on contamination of residential property buyers using contingent valuation surveys. *Journal of Real Estate Research,* 27(2), pp. 193-220, 2005.

[13] Simons, R., The effect of pipeline ruptures on non-contaminated residential easement holding property in Fairfax County. *The Appraisal Journal*, 67(3), pp. 255-63, 1999.

[14] Nelson, A. C., Genereux, J. and Genereux, M., Price effects of landfills on house values. *Land Economics,* 80(4), pp. 359-65, 1993.

[15] Greenberg, M., Craighill, P., Mayer, H., Zukin, C., & Wells, J., Brownfield redevelopment and affordable housing: a case study of New Jersey. *Housing Policy Debate*, 12(3), pp. 515-540, 2001.

Brownfield redevelopment by the private sector: market driven decision making

I. L. Whitman
The Whitman Companies, Inc., East Brunswick, N.J., USA

Abstract

Most successful Brownfield redevelopment in North America involves a partnership of interests between government and private developers. Much has been written of the need for "Public-Private Partnerships", as well as community involvement as contaminated Brownfield sites are recycled into economic productivity. Less well understood are the motivation and incentives for private investors and developers to take the risks necessary to succeed at Brownfield redevelopment, and the criteria used by the development community to determine when to do so.

The National Association of Industrial and Office Properties (NAIOP), of Herndon, Virginia, represent the largest and most experienced owners and developers of commercial real estate in North America. NAIOP is one of the few private sector organizations that has worked with USEPA and other public agencies to promote and enhance the understanding of the private sector opportunities and risks in Brownfield redevelopment. Interviews were conducted with NAIOP members and other Brownfield players regarding the market drivers that motivate private developers and their financial backers to invest their time and resources in Brownfield restoration and redevelopment, and the issues which stand in the way of such investment. Information developed from interviews and discussions with developers, and from NAIOP presentations at EPA's National Brownfields Conferences and from commercial real estate forums are cited and factored into the analysis of private sector forces that lead to success in Brownfield redevelopment.

WIT Transactions on Ecology and the Environment, Vol 94, © 2006 WIT Press
www.witpress.com, ISSN 1743-3541 (on-line)
doi:10.2495/BF060031

1 Introduction

In North America over the past decade, scholars and practitioners alike have worked to establish and promote a growing concept and practice geared to the revitalization and redevelopment of a myriad of Brownfield sites. Brownfields were first recognized for their environmental challenges, and subsequently for the many public policy and legal issues that were identified, and in time, largely solved. It is safe to say that as the technical, legal and regulatory elements of Brownfield redevelopment have been made manageable, the challenge going forward rests primarily with the private sector who provides the capital, skill and entrepreneurial energy necessary to actually transform substandard real estate into renewed resources with both private, and public benefits.

The workings of Brownfields redevelopment did not come smoothly or naturally to the real estate industry which cherishes predictability and freedom from risk. However, the prospect of new access to old land, and the difficulty of access to undeveloped land has led the real estate development and investment sector reluctantly but steadily into the Brownfield marketplace.

Real estate professionals associated with Brownfield activities agree that first and foremost, their decisions are governed by the real estate marketplace in which they operate.

1.1 National perspective

Thomas Bisaquino, president of the National Association of Industrial and Office Properties (NAIOP), a business association of major commercial developers in North America cites the lack of developable land in many of the older, more densely populated regions of the U.S. and Canada as providing the pressure to NAIOP's members to consider Brownfield sites in and near urban areas. Bisaqaquino indicated that approximately 20 of the 50 major U.S. real estate markets have evolved to where Brownfield redevelopment can compete with other development for investor capital and developer energy. These markets include most of the northeast corridor from Boston to Baltimore, Pittsburgh, Cleveland and Chicago.

Some of the demand for urban redevelopment is fueled by lifestyle demands among the aging "baby boomers" who are rediscovering urban centers and moving back to center cities from suburban locations where they grew their families. The market pressure is for so-called "lifestyle centers" where the population can live, work and play without continuing their dependence on automobile use. Infill development, selecting substandard parcels and groupings of properties in otherwise viable urban centers provides an attractive resource for developers to enhance sub value land, satisfy the lifestyle center demand, and make a reasonable profit. In every urban area, some of these infill properties will be environmentally suspect and warrant the full measure of Brownfield site due diligence, investigation and remediation.

1.2 Factors influencing market-based decisions

A wide variety of factors intersect with the real estate marketplace and the players in that marketplace that shape the individual decisions made by developers and investors over Brownfields. Factors that have been cited include risk and liability, return on investment, value creation, investor goals, developer profile, market changes, and government interaction.

2 Liability and risk

It can be said that successful Brownfield redevelopers hate liability and love risk.

2.1 Liability

Environmental liability was once the single most important factor in stifling Brownfield redevelopment. Under U.S. federal law, and state law as well, the liability for environmental contamination to properties is "joint and several", meaning each property owner in the chain of title back to when the contamination occurred is liable potentially for the entire cost of cleanup and environmental restoration. Under these conditions, investors and developers would not dare to consider acquiring property unless the full extent of environmental damage was known, and the property was well on its way toward complete remediation, a circumstance seldom applicable to under-utilized, devalued real estate. Attorneys would strongly advise clients not to venture into such shark-infested waters as a potential Brownfield site, and bankers would see to it that funds were not loaned for such purposes, lest they be somehow brought into the chain of liability themselves.

Susan Boyle, of the New Jersey Department of Environmental Protection (now on leave to the National Brownfield Association) assisted developers attracted to Brownfields by negotiating Prospective Purchasers Agreements on behalf of the state, and by providing developers with counseling on liability protection and reduction under the state's Brownfield statutes and regulations. Ms. Boyle has indicated that as developers become more sophisticated in such matters, the more confident they are in venturing into the Brownfields marketplace.

2.2 Risk

Dwight Stenseth, Managing Partner of Cherokee Investment Partners, Denver, Colorado, sees all prospective redevelopment properties as "risk adjusted", including environmental risks. Cherokee is exclusively a Brownfield redeveloper, using funds accumulated from investors who understand that all real estate carries some risk, and Brownfield sites may, on the average, carry a higher rate of risk. Stenseth indicates that to attract major investors, the return must be equal or greater than other real estate investment opportunities. Cherokee's investor base includes such establishment stalwarts as government pension funds, corporate pension funds and university endowments.

Several players in the real estate marketplace see Brownfields development as being about managing risk. Kenneth Cohen, owner/operator of Pantheon Properties of New York City, indicates that each investor and developer has a risk profile, which is related to anticipated return on investment. He believes that smaller, more entrepreneurial developers such as Pantheon must manage risk by having better knowledge of the real estate market, and substantial experience to know when environmental risk may become unmanageable. While insurance products are very valuable in helping a Brownfield developer manage risk, the insurance underwriters themselves succeed only through their knowledge and experience in dealing with the factors that create environmental risk.

Cohen and William Klein of Kay Realty Holdings, Ocean, New Jersey point out that all real estate investment and development entail risk, including the ultimate performance of the property in retaining tenants and producing an income stream necessary to provide the targeted return on investment.

3 Value and return

As entrepreneurial driven business people, developers and investors are motivated by the "bottom line", i.e., at the end of the day will the return on their efforts and investments provide the financial yield that could be obtained through alternative investments?

3.1 Value

Several of the developers interviewed, including Michael Perlman of Holualoa, Arizona of Tucson, Arizona, emphasize that Brownfield redevelopers "create value" by converting under-utilized properties to commercial assets that attract and retain tenants. Perlman, whose company acquires, restores and operates run-down, sub-value retail centers, does not look for Brownfield sites per se, but seeks to acquire property past its prime, physically inferior to newer developments, in areas where the upgrading of the facility will be accompanied by the demand for occupancy and the return of customers. Such facilities are often found in older suburban regions, where retail strip malls and shopping centers may be 35 to 55 years old, and environmental issues may include asbestos insulation, discharges from dry cleaning establishments and aging or no longer operating gasoline stations.

Cohen, of Pantheon Properties, is motivated by "fixing things and creating value". In the New York metropolitan area, Cohen finds that there are aging property owners holding sub-value properties, offering their properties at discount rates so that they can escape from the responsibilities of environmental liability and remediation. In some cases, these owners no longer have current knowledge of the commercial real estate market, a void which Cohen can fill and find the opportunity to restore the property to both its physical condition and its commercial value.

3.2 Return on investment

Return on investment is the standard by which most investments are assessed. In the development business, investors are the partners of developers. Whereas banks and other lenders may be involved at some stages in the process, particularly with smaller developers, capital is most often obtained through investor-partners. Investors set their sights on different levels of return depending on their obligations to the individuals or the institutions from whom the funds are raised. Stenseth indicates that Cherokee creates large scale investment funds, they raise money and are capital providers. Once created, the funds are invested in new redevelopment projects where the return on investment must exceed the obligations established to provide the investors with a competitive yield. Investors may be taken out of the project through a variety of complex financial structures that may range from high yield bonds, to secure loan repayments, to long term equity positions in the real estate market.

4 Owner-driven development

It is often believed that Brownfield redevelopment is driven largely by government authority seeking to restore and upgrade blighted areas, and by commercial developers motivated by a series of market forces. There are instances, however, where long-standing property owners are driving the process.

Donald Eisen, of Cushman and Wakefield of northern New Jersey, is a highly experienced broker in the commercial real estate market in the New York metropolitan area. Eisen has recently redefined his practice to focus on major corporate clients who have “mothballed” industrial properties no longer essential to their industrial operators. Eisen’s experience is that this class of property owner, quite substantial in certain locations, are looking not to redevelop their properties but to tap into the benefits of the real estate marketplace to bring about favorable sales of their property to potential redevelopers. In general, corporate owners are not expert in matters of real estate value, and Eisen and other senior brokers bring their expertise into this process on behalf of their clients.

Eisen indicates that his corporate clients are not motivated solely by the potential sale price of their property. He believes that corporate land owners desire to have the properties restored in a manner to create compatible uses with the community, so that their “brand name” property, which may have been a community icon for as long as a century or more, can be retired with a positive image that brings about a favorable reception from community leaders as the property is transformed from its present inactive use to its future identity.

5 Developer profile

Research shows that there are a wide variety of Brownfield redevelopers, and that the profile of the developers will largely determine the type of investment, the tolerance for risk, and the manner in which market forces drive their redevelopment objectives.

5.1 Size

In general, Brownfield redevelopers are smaller than the major commercial, industrial and residential developers whose prime motivation is to generate "product" in order to maintain a high volume of business and the continuous flow of revenue. However, in the regions of the U.S. cited by Bisaquino as "prime" Brownfield territory, major developers who never intended to get into Brownfields are now actively seeking out distressed properties because in some cases there are few other options. Throughout the northeast, major homebuilders are now developing substantial townhouse and condominium residences at Brownfield locations, because of the demand for and high value placed upon such locations. In some cases, the value of such "lifestyle" residences exceeds the value of comparable suburban prospects by several fold.

Perlman and Cohen point out that smaller developers need to be more entrepreneurial and selective in determining the projects that work for them. These "niche" players generally attract investors seeking higher returns who are more tolerant of the risks that may accompany real estate markets that center around Brownfield situations. Cohen indicates that the "big players" drive land prices up and return down, a financial reality that may satisfy their more conservative investors, but not the needs of the "niche players".

5.2 Niches

Smaller developers seek out and exploit niche situation because that is where their creativity, knowledge and experience can enable them to out-compete the larger companies whose financial expectations require them to look more broadly and globally at real estate markets. Bradley Tate, of Westrum Development Company located near Philadelphia, Pennsylvania, indicates that his company evolved into Brownfield development because, in their market area, that is where they found the most viable opportunities. Westrum, a niche player, "seeks redevelopment/revitalization opportunities in inner city and aging suburban towns and boroughs". In the greater Philadelphia area, they found that these opportunities came at "former industrial and manufacturing facilities with significant environmental challenges". Among the forces that will drive a successful deal for Westrum are "strong market conditions" and "high profit potential".

5.3 Knowledge and experience

Successful Brownfield developers believe that their particular experience and knowledge give them an important advantage in their marketplace. In some cases, knowledge of environmental issues is cited as an advantage, but, in truth, this specialized knowledge can be obtained externally by working with experienced environmental consultants who understand the complexities of remediation in the Brownfield marketplace. More importantly, it is *real estate expertise* that is invaluable. Knowledge of tenants, government approvals and land use restrictions, and the financial marketplace are most essential.

It is no accident that the interviews conducted in the preparation of this paper were with *successful* players in the Brownfields marketplace, whose expertise has sustained them throughout their business enterprises. Obviously, the unsuccessful players, who are no longer participating in Brownfields redevelopment, are neither available nor in some cases aware that lack of marketplace expertise may have contributed to their failures.

6 Conclusions

Although conceived as a means of remediating contaminated property and reducing hazards to the environment, Brownfield redevelopment, in the final analysis, is a real estate concept that will succeed or fail, in each case, based on real estate principles. The majority of Brownfield sites restored, and value created will be done so at the hands of the private sector.

It is clear that entrepreneurial redevelopers hold the keys to the Brownfield domain, and that they are largely driven by conventional real estate market forces, forces that are likely to rise and fall with the tide of the real estate marketplace.

Section 2
Environmental assessment

Environmental accounting of a waste management system: outcomes from the emergy analysis

C. Granai, R. M. Pulselli & R. Ridolfi
Department of Chemical and Biosystems Sciences,
University of Siena, Italy

Abstract

This paper shows an application of the emergy analysis to the waste management system in the Province of Cagliari, Italy. It aims to consider the importance of waste management within a territorial system by assessing the environmental cost and gain due, directly and indirectly, to waste management processes. It also aims to investigate different steps of this process looking for potential critical points and available strategies for a sustainable waste management in its three phases of waste gathering, waste treatment and waste disposing. A landfill and an incinerator, located within the Province, are studied in order to consider their own environmental efficiency and their role in the waste management process. The emergy method (spelt with an "m") is based on the assessment of all the inputs that supply the waste management system and its processes (emergy means energy memory). The expected outcomes are in the form of a comprehensive balance that shows the general behaviour of the system and its parts. Through the emergy synthesis, flows of energy and matter that are used within the process of waste management in the Province of Cagliari, are assessed by a common unit, namely the solar energy joule, and compared to each other. This is performed in order to evaluate how an integrated system including incinerator and dumps works in order to make choices and to achieve a sustainable waste management plan.

1 Introduction

Global environmental conditions are seriously threatened by man and its growing production and consumption rates, especially in western countries. One aspect of

WIT Transactions on Ecology and the Environment, Vol 94, © 2006 WIT Press
www.witpress.com, ISSN 1743-3541 (on-line)
doi:10.2495/BF060041

the problem is the fact that waste production is greater than the capacity of ecosystems to absorb them. The effects of the accumulation of discharged matter, substances, water, etc., are somewhere evident and the consequences (for instance, the harmful effects on the population) must induce to implement rapid and clear political decisions. For example, disposing this huge amount of wastes in a more efficient (from an economic as well as energetic viewpoint) and less detrimental way would mean managing a "new" resource in compliance to Herman Daly's principles of sustainable development [1].

In a thermodynamic context, waste is defined as an entropic reject. However, from thermodynamic theory we know that the ultimate and degraded form of energy is heat, while any material consistence of tailings is actually a resource that can and must be used in some production process. The aim of a public planner in this field, at local or national level, should consist in selecting the most suitable and efficient procedures and technologies in order to appreciate this "new" resource.

The recycle of matter and the recover of energy are arduous procedures that need careful planning and high initial investment without immediate economic advantages. This kind of projects would provide tangible results in the long run.

This paper presents the application of an environmental accounting system, namely the emergy evaluation, to the waste management system of the Province of Cagliari, with the purpose of evaluating, both quantitatively and qualitatively, some crucial points in order to orient public policy in this field. The main elements of the analysis are the dimension of the system, the energy and matter involved within; the importance of this sector within a territorial sustainability evaluation of the same Province. The analysis enables to highlight the environmental problems related to each phase of the system: gathering, management and disposal of wastes.

2 Emergy analysis of a waste gathering system

The emergy analysis (spelt with an "m") measures the main resource flows that sustain a territorial system and its population. This accounting method converts different inputs, such as human labor, trucks, fuels, chemicals, to a common unit (the equivalent solar energy) [2, 3]; it describes the environmental state of a region with respect to life standard and activities of local population.

Assigning an emergy content to waste means that waste has a certain potential value and does not represent a state of maximum entropy; it is still a resource for the system. Since residues of an ecosystem are the feedstock for another, even Nature does not conceive the concept of waste [4].

However, through the work of Herman Daly, one of the father of Sustainable Development and Ecological Economics, the classic and neoclassic economy, based on the parameters of labor and capital produced by men, have been reloaded by a third parameter, "the natural capital".

This paper presents the emergy analysis of waste as an attempt to introduce the "natural capital" for the evaluation of a waste management system, not only assessing monetary costs and waste quantity. In particular we want:

- to measure the value of collected materials in terms of *solar emergy joule* in order to estimate the amount of *natural capital* lost in the form of waste;
- to promote the idea that waste is still a resource and not only something to dispose;
- to prove the environmental advantage of recovering waste trough an integrated management;
- to propose a hierarchy of possible waste treatments pursuing a high efficiency of the waste management system;
- to suggest new routes for a sustainable waste management to Public Administration.

The emergy content of each material category is given multiplying mass quantities by the transformity.

Table 1: Differentiated and undifferentiated waste production in the Province of Cagliari.

Materials	PROVINCE OF CAGLIARI		
	Differentiated	**Undifferentiated**	**Total**
YEAR 2001	**(t)**	**(t)**	**(t)**
TOTAL	**10,10**	**383,00**	**393,000**
ORGANIC	461	149,000	150,000
PAPER	2,78	119,000	121,000
GLASS	2,14	30,600	32,800
PLASTICS	359	38,300	38,600
ALUMINIUM	0.31	15,300	15,300
BULKIES	4,32	-	4,320
OTHERS	62.7	30,600	30,700

Table 2: Emergy flows of differentiated and undifferentiated waste.

Materials	Solar Transformity	PROVINCE OF CAGLIARI		
		Differentiated	**Undifferentiated**	**Total**
YEAR 2001	**(sej/g)**	**(sej)**	**(sej)**	**(sej)**
EMERGY		**1.34E+21**	**2.95E+19**	**1.37E+21**
ORGANIC	3.77E+09	5.63E+20	1.74E+18	5.64E+20
PAPER	3.15E+09	3.74E+20	8.75E+18	3.82E+20
GLASS	8.40E+08	2.57E+19	1.80E+18	2.75E+19
PLASTICS	3.80E+08	1.45E+19	1.36E+17	1.47E+19
ALUMINIUM	1.60E+10	2.45E+20	4.94E+15	2.45E+20
BULKIES	3.90E+09	-	1.69E+19	1.69E+19
OTHERS	3.77E+09	1.15E+20	2.36E+17	1.16E+20

Mass quantities give the waste system size. The transformity of each material is a specific intensive quantity that depends on its production process. The transformity is therefore the emergy necessary to obtain one unit of a material. Once we have obtained the emergy of each material (multiplying mass quantities by the transformity), we can sum the emergy content of each item; a measure of the total emergy flow corresponding to waste is then achieved (Table 2). In the Province of Cagliari, the emergy flow of waste is 1,37 x 10^{21} sej.

In Figure 1 is shown a graph with the total amounts of emergy flows due to different materials from waste gathering. Organic waste and paper are the most relevant, respectively 41% and 28% of the total emergy flow. Aluminium is 18% due to the high value of the transformity.

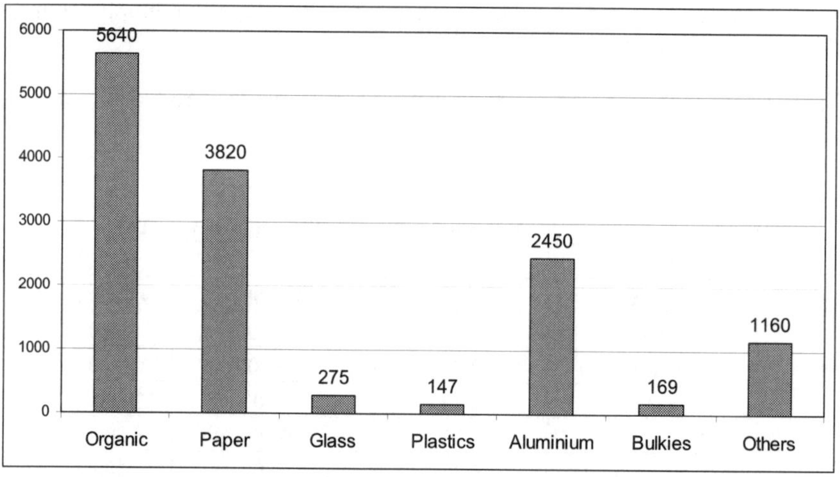

Figure 1: Emergy flows of waste production in the Province of Cagliari.

3 Emergy analysis of waste treatment and disposing system

The emergy analysis applied to a waste treatment and disposing system follows a *life cycle analysis* that provide a schematic description of the process throughout the product's life, *from cradle to grave*. The waste treatment system has been divided into three different phases: waste gathering (it includes the transportation of wastes from the city's containers to the waste treatment plant), waste treatment (the real process of waste management), and residues treatment (safe deposal of residues) [5]. All inputs, necessary to feed the process, are aggregated per phase and converted in solar emergy units. The aim of this analysis is not to measure the value of waste, in terms of natural capital, but to evaluate the environmental efficiency of a the treatment system in the Cagliari province, and also to evaluate energy recovering processes from waste [6]. Consequently, resources necessary to treat wastes have been accounted, from gathering to the final disposal. From a thermodynamic viewpoint, the best process is the one able to join low emergy investments with an environmental compatible management and with an efficient recovery of materials and energy.

Table 3: Compared emergy flows of a landfill and an incinerator in the Cagliari province. Emergy investment per ton of treated waste and per MWh of electric production.

TYPE of TREATMENT	Gathering	Treatment	Residues	**Total**	**Investment**	**Investment**
	sej	sej	sej	**sej**	**sej/t**	**sej/MWh**
LANDFILL (Serdiana)	1.14E+19	1.38E+19	3.47E+16	**2.52E+19**	**1.92E+14**	**1.54E+10**
INCINERATOR (casic)	1.20E+19	2.24E+19	8.64E+18	**4.30E+19**	**2.07E+14**	**5.49E+09**
TOTAL	2.34E+19	3.62E+19	8.67E+18	**6.82E+19**	-	-

Table 4: Emergy analysis of the landfill (Serdiana).

	Item	**Units**	**Quantity** (units/year)	**Solar Transformity** (sej/units)	**Solar Emergy** (sej/year)	**Emergetic Investment** (sej/t/year)
COLLECTION						
1	Collected wastes	t	1,31E+05			
2	Collection cost	€	8,14E+06	1,40E+12	**1,14E+19**	**8,68E+13**
TREATMENT						
3	Plant cost	€	3,82E+05	1,40E+12	5,35E+17	4,07E+12
4	Human work	J	1,10E+10	7,38E+06	8,15E+16	6,20E+11
5	Material	g	2,34E+10	5,00E+08	1,17E+19	8,91E+13
6	Machinery	g	1,19E+07	6,70E+09	7,98E+16	6,08E+11
7	Fuels	J	1,14E+13	1,21E+05	1,38E+18	1,05E+13
8	Electricity	J	4,87E+05	2,00E+05	9,73E+10	7,41E+05
9	Water	g	1,50E+09	2,03E+05	3,05E+14	2,32E+09
					1,38E+19	**1,05E+14**
RESIDUAL MATTER TREATMENT						
10	Percolate disposal cost	€	2,48E+04	1,40E+12	**3,47E+16**	**2,64E+11**

Electric Energy Production	4,50E+13	(J/year)
	1,25E+04	(MWh/ year)
Total Solar Emergy (sum of items from 1 to 10)	2,52E+19	(sej/year)
Phase 1: Collection (item 2)	1,14E+19	(sej/year)
Phase 2: Treatment (sum of items from 3 to 9)	1,38E+19	(sej/year)
Phase 3: Treatment of residuals (item 10)	3,47E+16	(sej/year)
Emergy investment per ton of waste	1,92E+14	(sej/t/year)
Emergy investment per MWh of energy	1,54E+10	(sej/t/MWh/year)

3.1 Managed landfill (Serdiana)

The plant located in Serdiana municipality is a landfill of first category. It has a total authorized volume of 425,000 m^3 with a residual useful capacity of 20,000 m^3 in 2001. The general cost of the plant during the year 2001 is estimated to be 740,000.00 euros. The plant has been constructed in 1997. Annual average working hours have been estimated in 21,100 per 24 working units. Waste buried in the landfill during the year 2001 are 131,000, an average of 250-300 t/g. leachate is disposed in a depuration plant, when it is not recycled in the landfill. The average cost of disposal is 31 €/m^3. The average biogas production is 1,500 Nm^3/h. Electricity production is equal to 12,500,000 kWh/year. 13,000 m^3 of inert material are used to bury waste.

From the emergy analysis of Serdiana plant we have obtained the following results:

- The 55% of all the emergy used is due to the treatment phase.
- The remaining 45% is mostly relative to waste gathering. The emergy cost of the disposal is almost negligible. This result depends on the low quantity of leachate treated.
- The 46% of the total emergy flow is due to materials use for the landfill cultivation. The other two meaningful items are plant cost and fuel use, which represents about the 7-8% of the total emergy.
- The emergy investment per ton of waste treated is 1.92x10^{14} sej/year.

The emergy investment per MWh of electricity produced is 1.54x10^{10} sej/t/year.

Table 5: Emergy analysis of the incinerator (Casic - Macchiareddu).

Item	Units	Quantity (units/year)	Solar Transformity (sej/units)	Solar Emergy (sej/year)	Emergetic Investment (sej/t/year)
COLLECTION					
1 Collected wastes	t	1,66E+05			
2 Collection cost	€	8,57E+06	1,40E+12	**1,20E+19**	**7,23E+13**
TREATMENT					
3 Human work	J	4,13E+11	7,38E+06	3,04E+18	1,83E+13
4 Electricity	J	6,54E+13	2,00E+05	1,31E+19	7,88E+13
5 Plant cost	€	3,62E+06	1,40E+12	5,06E+18	3,05E+13
6 Fuels	J	5,11E+12	1,21E+05	6,18E+17	3,72E+12
7 Water	g	3,84E+11	2,03E+05	7,80E+16	4,70E+11
8 Chemicals	€	3,36E+05	1,40E+12	4,70E+17	2,83E+12
				2,24E+19	**1,35E+14**
RESIDUAL MATTER TREATMENT					
9 Slag and ash	g	3,47E+10	2,49E+08	**8,64E+18**	**5,20E+13**
Electric Energy Production	J	1,36E+14			
	MWh	3,77E+04			
Total Solar Emergy (sum of items from 1 to 8)		4,30E+19	(sej/anno)		
Phase 1: Collection (item 2)		1,20E+19	(sej/anno)		
Phase 2: Treatment (sum of items from 3 to 8)		2,24E+19	(sej/anno)		
Phase 3: Residual Matter Treatment (Phase 9)		8,64E+18	(sej/anno)		
Emergy investment per ton of waste		2,07E+14	sej/t		
Emergy investment per MWh of energy		5,49E+09	sej/MWh/t		

3.2 Incinerator (Macchiareddu)

The incinerator located in Macchiareddu is managed by Tecnocasic. It has three burners: two dedicated to MSW and RDF (Refuse Derived Fuel) combustion and one to the incineration of industrial and medical wastes. The cost of the plant that works since 1995 is about 30 million euros. In 2001, the amount of waste incinerated was 136,317 tons, while the amount landfilled is 49,191 tons.

The electricity production, in 2001, from the three burners is about 37,676,100 kWh excluding a consumption of 18,174,450 kWh. The incineration process produces 30,593 tons of drosses and 4,102 tons of ashes.

The emergy analysis of the incinerator shows the following results:

- 52% of the total emergy is due to the treatment phase.
- Waste gathering represents the 28% while the final disposal is about the 20% of the total emergy use.
- The most important items are the electricity and the plant use that represent, respectively, the 31% and the 12% of the total emergy use.
- The emergy investment per tons of waste treated is 2.07×10^{14} sej/year.
- The emergy investment per MWh of electricity produced is 5.49×10^{9} sej/year.

4 Conclusion

The analysis of an integrated waste management system needs a holistic approach to evaluate various steps of the process in a unique balance from waste gathering to waste treatment and disposing. Results highlight the importance of strategic choices addressed to decrease waste production, to increase material recycling and to promote energy production from wastes as a final step, before waste disposing (the ultimate step).

This analysis intends to support choice making process through an environmental accounting method enhancing the idea that the environment is not just an economic *externality* but a form of capital, namely *natural capital.*

Results obtained enable to evaluate the environmental weight of waste in terms of solar energy spent or saved, and also the environmental efficiency of two principal processing systems (dumping and incinerator).

The integrated managing system is based on a set of processes for waste management.

Emergy analysis is here applied to evaluate energy and material flows involved in the different processes of waste management.

- ✓ The total emergy flow of waste is 1.37×10^{21} sej; this represents a resource from which to obtain the maximum possible energy outcome.
- ✓ Emergy use by the waste management system in various steps (considering a landfill and an incinerator), in the Cagliari Province is 6.83×10^{19} sej.

Although the waste management system consumes relevant amounts of resources. This investment, in terms of natural capital, is desirable because of the need to manage waste in a modern society and also because of the opportunity to produce energy, in alternative to fossil fuels.

- ✓ Electric energy production from waste in the Province of Cagliari is equivalent to 1.81×10^{14} J, corresponding to 3.61×10^{19} sej.
- ✓ The landfill in Serdiana has an emergy investment of 1.54×10^{10} sej per MWh of produced energy; the incinerator in Macchiareddu has a lower value, 5.59×10^{9} sej per MWh.

Dumping needs for a high investment but it extends negative effects due to waste emissions in time. A landfill is like a storage in which waste is stocked generating related problems.

An incinerator, opportunely managed and addressed to produce energy, processes a higher amount of waste decreasing their volume. Energy production from waste could be an environmental efficient process since it saves energy from fossil fuels, and it seems to be a good way to compensate emergy investment.

Anyway, practices for waste production decreasing, waste reusing, material recycling from differentiated waste gathering must be the first solutions to adopt to restrain the waste problem.

References

[1] Daly, H.E. Toward some operational principles of sustainable development. *Ecological Economics*, **2**, pp. 1-6, 1990.

[2] Odum, H.T. Self organisation, transformity and information. *Science*, 242, pp. 1132-1139, 1988.

[3] Odum. H.T. *Environmental Accounting, Emergy and Environmental Decision Making*. Wiley & Sons, New York, 1996.

[4] Riva, A. and Tiezzi, E. 1997. I rifiuti come risorsa sostenibile, in *Gea Speciale* (Rimini, Italy): 11:18.

[5] Marchettini, N., R. Ridolfi, E. Tiezzi, 2002. EMergy assessment of an integrated municipal solid waste management. *Waste Management and the Environment* (C.A. Brebbia, D. Almorza, D. Sale & V. Popov eds.), WIT Press, Southampton, 575-83, 2002.

[6] Marchettini, N., Ridolfi R., Rustici M.. An environmental analysis for comparing waste management options and strategies. *Waste Management, in press*, 2006.

What constitutes success in brownfield redevelopment? A review

T. Silverthorne
Social Research Group, Forest Research, UK
Centre for Environmental Strategy, University of Surrey, UK

Abstract

This review identifies and examines the different definitions of success that appear in the literature on brownfield redevelopment and discusses perceptions of: greenspace creation; economic issues; sustainability and success models. It concludes with some insights into what is currently perceived to constitute success in brownfield redevelopment. Brownfield redevelopment is generally acknowledged as one of the principal factors in ensuring that development is sustainable, but there is neither a benchmark standard nor a list of criteria by which success can be defined. In many redevelopments that have been judged, primarily by the developers, to be successful there is little correlation between the criteria on which the success has been claimed and the social or economic wellbeing of the local residents and success may in fact occur at their expense. Success is usually measured against the original objectives of the project where the focus is often more towards economic factors rather than social and environmental factors. All of these variables make it very difficult to generically quantify success in brownfield redevelopment.
Keywords: evaluation, success, brownfield redevelopment, sustainable development, urban greening, green space creation.

1 Introduction

This literature review was carried out using internet searches and on-line databases such as Web of Knowledge. The review identifies and examines the different definitions of success for brownfield redevelopment that appear in the literature. The search focused on the published literature contained in journals, papers and reports of the past ten years predominately from Europe and North

WIT Transactions on Ecology and the Environment, Vol 94, © 2006 WIT Press
www.witpress.com, ISSN 1743-3541 (on-line)
doi:10.2495/BF060051

America. Search terms included: 'brownfield'; 'evaluation'; 'redevelopment'; 'success'; 'urban greening'.

In the UK, there have been many different definition of 'brownfield' reflecting the perspectives and practices of a wide range of users. The most commonly accepted definition of 'Brownfield' by planners and other stakeholders in the UK is '*any land or premises which has previously been used or developed and is not currently fully in use, although it may be partially occupied or utilised. It may also be vacant, derelict or contaminated. Therefore a brownfield site is not available for immediate use without intervention*' Alker *et al.* [1, p. 64]. This crucially differs from the American definition because the site does not have to be contaminated to be classified as brownfield. Therefore, although contamination may not be present, a brownfield site is not necessarily available for immediate use without intervention. In Britain, The National Land Use Database (NLUD) states that there is currently 66,000ha of brownfield land (EP [2]) and suggests that 6,500ha of previously developed land becomes classed as brownfield every year (ODPM [3]).

An evaluation of the success of brownfield redevelopment projects requires a benchmark of success against which the success of individual projects can be measured. However, there is no overarching definition of success contained in the literature and the majority of evaluations are on a very site specific basis. Therefore, in view of the increasing emphasis on the need for brownfield redevelopment and the large sums of money now being invested in these projects, it is very important that there should be a broader definition of success. This review highlights the need for such a definition and suggests factors that should be included in such a definition.

England is the most densely populated country in Europe and is the fourth most densely populated country on the Earth (EP [2]) with four out of five inhabitants living in urban areas (DETR [4]). By 2016 it is estimated that 1.1 million new houses will be required in the south-east (ODPM [5]) to accommodate an expanding population, changes in patterns of household occupancy such as increasing rates of single parent and single person households and the effects of increased life expectancy.

In order to meet government targets for the building of infrastructure and new homes, Carey [6] concludes that brownfield sites will need to be rapidly brought into use. According to the Council for the Protection of Rural England (CPRE), there is now a greater need than ever for brownfield redevelopment due to the increasing threat to the greenbelt by current Government policies. In 2005 there were approximately 160,000 new homes built in the south-east (CPRE [7]), the Sustainable Communities Plan aims for 200,000 new homes in the south-east every year by 2016 and, according to the CPRE, a series of greenbelt boundary reviews are planned (GP [8]).

Brownfield redevelopment is generally acknowledged as one of the principal factors in ensuring that development is sustainable (Grimski and Ferber [9]). According to the OECD [10], brownfield redevelopment is '*part of a larger effort to address land use problems throughout urban regions, the better to control urban sprawl, improve social cohesion and create conditions favourable to local economic development*' (Moore [11, p. 326]). The problem is that brownfield redevelopment is often viewed as having either no, or even a negative, economic value [12–14]. In the past, in the face of increasingly powerful economic pressures, there has been a tendency to define the success of brownfield redevelopment schemes in economic terms to justify the expense and effort involved. Therefore it is now of vital importance when formulating a definition of success of redevelopment projects to include factors relating to the success of the social objectives with regard to the original stakeholders and the local communities.

2 Definitions of success

The main problem with defining success is that while appearing simple, it is in fact a concept that has numerous dimensions and there is not one comprehensive model or formula within existing frameworks (Layne [15]).

A definition of what constitutes success contained in a report for the Department of the Environment, Transport and the Regions (DETR) is 'if an area no longer experienced economic, social housing and environmental problems – except those displayed on average throughout England' (DLEUC [16, p. 59]). This implies that for the government success is just an absence of failure.

The way success is generally judged is on a site specific basis and simply involves an appraisal of whether or not the original objectives of the stakeholders have been achieved. For example, at the Fairfield Ecological Industrial Park in Baltimore (USA) more than 13,000 acres were zoned for industrial development. The notion of what constitutes success from this site is derived purely from the original intention for industrial ecology and does not include a wider concept for a standard of success (SGN [17]).

Success can also be measured by: environmental remediation; civil infrastructure renewal; job creation; tax-based development; economic development and neighbourhood revitalisation (Amekudzi and Fomunung [18, p. 211]). Although, it is worth noting that the measurement of these indicators can be very difficult and time consuming and the results are often disputed because the methodologies used to collect the data are often inappropriate. Also, the wide range of variables that can influence these factors make it difficult to prove a clear cause and effect relationship.

A more simplistic approach to measuring success is the view that any redevelopment of a brownfield site must be an improvement and therefore a success due to the inherent problems and costs associated with unimproved brownfield sites. These costs can be classified as economic, social and environmental (Burnham-Howard [19]). An unimproved brownfield site can attract a great deal of anti-social and criminal behaviour including arson, fly-tipping and vandalism.

Brownfield redevelopment has often been seen as a success in terms of both improving environmental quality and reducing the demand for greenfield development. Indeed, the US Council of Environmental Quality stated that 'every acre of reclaimed brownfields saves 4.5 acres of greenspace' from being redeveloped (WHCEQ [20]).

3 Green space creation

One definition of success by planners, applied geographers and economic development practitioners is to 'create liveable spaces that enrich the daily lives of residents' (Gatrell and Jensen [12, p. 331]). Within the US, there has been a move towards alternative economic development with 'smart growth' economic plans. These try to balance the often competing interests of the economy, the environment and quality of life by focusing on the overall liveability of the locality and the importance of a sense of place (English [21]). Within this school of thought, urban forestry is seen as encapsulating successful brownfield redevelopment due to its contribution to the quality of life in the locality. Urban forestry can be coupled with economic development and a community's smart growth strategy as studies from Florida have shown.

Within the UK, there has been a growing trend for greening, more so than in the US. For instance, between 1988 and 1993, over 19% of brownfield redevelopments were converted into green spaces, which was more than any other end use (DETR [22]), whereas in the US, the emphasis has been on economic benefits (De Sousa [23]). However, this is not to say that economics are not a consideration in the UK. The North West Development Agency views the greening of the north-west of England as bringing economic benefits through investment in an attractive area and subsequent relocation of business.

When it comes to redevelopment in the form of urban greening, involving 'all public and private urban spaces that are not built or otherwise fully impervious' (Jensen *et al.* [24, p. 118]), there is a growing definition of success in the literature that states that green redevelopment should involve a multi-functional view of development that encompasses economic, environmental, recreational and aesthetic issues and involve communities in the decision making process.

Urban green space is becoming an increasingly popular option for brownfield redevelopment with 33 million inhabitants making an estimated 2.5 billion trips each year to these areas in the UK (Dunnett *et al.* [25]). Money is now being made available for urban green space development on brownfield land by the Heritage Lottery Urban Parks Fund (ODPM [26]). These areas can be judged to be a success by looking at their utilisation for social activities. Dunnett *et al.* [25] listed the following activities: sporting activities; quiet time; walking; attending events; social activities and passive activities such as reading or sitting.

From a social point of view, one of the criteria for success is its aesthetic appeal and characteristics. Although there are some attempts in the published literature to assess the beauty of forests and green areas [27, 28], this is notoriously difficult to assess and quantify.

Studies have shown that the success of brownfield redevelopment is not solely due to environmental factors and that success cannot be achieved by just addressing the environmental issues. This was the conclusion of a study that was carried out in the US by Lang and McNeil [29]. The study concluded that the differences between successful and non-successful redevelopment sites were not based on site specific criteria such as size; distance to an airport or city centre; rail access or frontage onto water but were dependant on: time to occupancy; land use; preservation / creation of greenspace; relative cost of remediation; availability of incentives; integration of the property into the neighbourhood and the willingness of the banks and political forces to find co-operative solutions (Lang and McNeil [29, p. 107]).

The study also concluded that due to the importance of land-use and time to occupancy, when assessing the success of a redevelopment project, there is a need to focus on the two primary stakeholder groups: the community, who need to be assured that the development is in tune with the social and economic needs of the community; and the developer / investor who looks for a return on their investment (Lang and McNeil [29]).

From an economic perspective, it is more difficult to assess the success of a brownfield greening project such as urban forestry because urban green areas do not usually generate income and are generally perceived as public services without any economic value [12, 13]. However, there are likely to be strong benefits to the surrounding area in the form of higher property values. Protagonists argue that environmental improvements such as the presence of trees and forested areas create better living spaces which will also have the effect of improving the quality of economic development and an increased property value [30, 31]. This trend can also be seen in the UK. Thames Barrier Park in London is a free public amenity and so does not make much money, but it led to executive flats being built in the immediate surrounding area (Holden [32]). Tyrvainen and Miettinen [33] observed that there is also a statistically significant correlation between closeness to green space and the value of property with a 5.9% decrease in market price with every km away from a wooded area. There is also a rise in the quality of life indicators (Gatrell and Jensen [12]). This form of redevelopment ultimately enhances the local system (Anderson and Cordell [30]), thus making for a successful redevelopment.

4 Economic issues

One of the main disincentives for developers facing brownfield redevelopment is that the focus is all too often on economic benefit; however, the benefits to the stakeholders are frequently intangible, with the implication that any other benefits are secondary and a possible positive side-effect. This point of view is expressed by Burnham-Howard [19, p. 213] who stated 'Each brownfield site represents untapped potential for economic development in the form of community benefits in jobs and tax revenue'. Brownfield greening can help the local economy because it is now generally accepted that 'what tends to attract higher economic functions, besides the quality of the labour force, is the quality of the environment' (Zetter [34 p. 57]).

Because brownfields are historically and predominately viewed in economic terms (although this is beginning to change, albeit slowly) success is often judged within this narrow category. Howland [35, p. 2] concluded that two brownfield redevelopment sites in Baltimore (US) were a success because they have 'profitable enterprises now operating on both' with just three years between the purchase of land to a profitable operation. This economic focus is due to the fact that in the US, brownfields are viewed in the main as 'a currently unrealised asset' (Lang and McNeil [36]) or as 'under utilised' (De Sousa [23, p. 181]). Similarly, in the UK the prevailing view is of brownfield land as 'wasting assets' (UTF [37]) and this has given rise to the dominant view that redevelopment is a way to bring this land into beneficial use. Therefore any redevelopment can be classed as a success because any alternative is better than a derelict brownfield site with its association with socially undesirable activities.

If success is only measured from an economic perspective, it is relatively easy to achieve a successful outcome. There have been numerous success stories from the US over the last 10 years with idle and under-utilised contaminated properties now housing a variety of economic activities, both single and mixed use, such as leisure complexes, urban villages, commercial estates and industrial estates. It is, however, unclear whether these developments have helped to revitalise the surrounding area by contributing to a more sustainable community. One of the reasons for this is that the usual gross metrics such as job creation, often fail to distinguish between jobs transferred from other industries and new job creation, nor do they take into account the more social aspects of the community such as quality of life, social cohesion and environmental justice.

5 Sustainability

The redevelopment of brownfield land can be viewed as successful if it yields substantial results in the three pillars of sustainable development: environmental, social and economic (Andrew [38]) and so the success of redevelopment has the potential to solve multiple problems through a single activity (Strother [39]).

A major difficulty with judging the sustainability success of brownfield redevelopment based on sustainable development, especially the social aspect, is the problem of measurability. Some of the aspects of sustainability, especially the social aspects are difficult to quantify, such as the aesthetic values, community interaction and feelings of safety. An additional problem is the fact that environmental and economic objectives are often judged to be incompatible with each other.

Mixed-use developments have been seen as successful because of their potential to combine the elements of sustainable development. It can be a stand alone development such as an urban village, e.g. the Hulme area in Manchester (EP [40]), or where smaller mixed use projects are part of a larger regeneration drive. One successful example can be found in Norrkopping, Sweden (CLARINET [41]), another at the Salts Mill in Bradford, where a former textile mill is now a mixed use development incorporating an IT research facility, retail, cultural and commercial outlets with over 750 people employed on the site (Alker and McDonald [42]). According to Alker and McDonald [42], the success of this type of project is a result of the integration of the community and the other stakeholders who all assessed aspects of the proposed redevelopment.

In the current political climate, due in part to the growing trend of ecological modernisation (Buttel [43]), developers are adopting a greener approach and so newer developments are offering advantages and opportunities for wildlife where ten years ago, none would have been forthcoming. Thus offering another opportunity for redevelopment to be considered a success.

6 Success models

There is a lack of tools for dealing with the assessment of brownfield redevelopment. The majority of the literature deals with anecdotal and qualitative studies and there are very few systematic studies that show what is successful and what is not and, more importantly, how to define a successful brownfield redevelopment. Nijkamp *et al.* [14] have developed a model from Dutch case studies that uses a meta-analytic approach for determining success when redeveloping contaminated brownfield sites. However, this looks more at the success of the process of redevelopment rather than the final outcome including problems with financing, jurisdictional responsibility and regulatory compliance.

More recently, Lang and McNeil [29] have developed a logit model which utilises data from 75 brownfield sites in the US and analyse the data to distinguish between the successful and the unsuccessful. They state that this can be used to estimate the probability of success. When making the model, they concluded that successful sites: incorporated more greenspace into the development plan; were more likely to take advantage of existing infrastructure; were more likely to have financial incentives available to the developer; were better integrated into the neighbourhood and will have a positive effect on local businesses and considered the future use of the property when establishing environmental cleanup levels (Lang and McNeil [29, p. 110]).

7 What constitutes success?

There are many different definitions concerning the success of brownfield redevelopment. There does not seem to be one bench mark standard. Bodies such as the US Environment Protection Agency (EPA) and the Office of the Deputy Prime Minister (ODPM) do not have a list of criteria for which success can be defined. Instead, success is measured against the original intentions of the redevelopment, which is broad ranging and on a site specific basis. The majority of the focus though has been on the economic success of a site with relatively little done to assess the social success of the site. This is because traditionally, the reason for regeneration is an economic one.

There is also variation due to the different types of redevelopment and the ultimate end-use of the sites. Different countries have different ideas as to what type of greenspace is wanted (Grimski and Ferber [9]). All of these variables make it very difficult to generically quantify success for brownfield redevelopment.

Perhaps one of the most important questions, and the most overlooked, when addressing the success of redevelopment projects is the question 'success for whom?' There have been a number of authors who have observed that the successful redevelopment does not always correlate with the social or economic wellbeing of the local residents, and may in fact occur at their expense [15, 44].

Environmental improvement is not usually perceived as a criterion for success. In a study conducted by Wernstedt *et al.* [45, p. 10], more than 80% of the respondents of a public sector survey thought that a more efficient use of public infrastructure, the increase in tax revenue, neighbourhood revitalisation and the creation of jobs were the main reasons for brownfield redevelopment and it is on this basis that they would be judged a success. The attention and priority appears to be on social and economic success factors rather than environmental.

There is a growing debate in the literature that states that the involvement of all stakeholders (people in the community, the developer, local government and regulators) from the beginning is essential for successful brownfield regeneration and the sites integration into the surrounding community (Stenner *et al.* [46]). Villavaso *et al.* [47] argue that local communities need the tools to 'assess, cleanup and reuse contaminated [brownfield] properties' with an integration between the local community and the possible developers and they go on to suggest how this can be achieved for successful redevelopment.

There is now wider recognition that successful brownfield redevelopment should encompass sustainable development with ecological sustainability; social equity; and economic efficiency with clearly defined long-term objectives so that the expectations of stakeholders remain constant (Agrawal [48]). However, Pagdee *et al.* [49] argues that a proportionate combination of these three pillars may not exist owing to nearly all of the casual elements being subject to change. It is clear that there is no one definition of success as it depends on what the objectives and aims of the site were, one's personal perspective and interests and on definitions of success which may vary amongst the stakeholders of the same redevelopment project. There is, however, a clear theme emerging that any definition of success must ensure that there is meaningful consultation by the developers with the local community; potential stakeholders including user groups; the project is sustainable for the long-term; improves the local environment and addresses the needs of the local community.

Acknowledgements

The author is registered on an Engineering Doctorate Programme at the University of Surrey and gratefully acknowledges the support of the EPSRC and Forest Research. The author would also like to thank Dr. Liz O'Brien at Forest Research and Dr. Walter Wehrmeyer, Dr. Kate Burningham at the University of Surrey.

References

[1] Alker, S., Joy, V., Roberts, P. & Smith, N., The definition of brownfield, *Journal of Environmental Planning and Management*, **43(1)**, pp. 49-69, 2000.

[2] EP [English Partnerships], *Towards a national brownfield strategy,* English Partnerships: London, 2003.

[3] ODPM [Office of the Deputy Prime Minister], Landfill tax and the development of brownfield land. http://www.odpm.gov.uk/stellent /groups/odpm_urbanpolicy/documents/pdf/odpm_urbpol_pdf_608149.pdf

[4] DETR [Department of the Environment, Transport & the Regions], *Contaminated land: Implication of part IIA of the Environmental Protection Act 1990*, HMSO: London, 2000.

[5] ODPM [Office of the Deputy Prime Minister], Delivering growth in the Thames Gateway and the growth areas. www.odpm.gov.uk/stellent /groups/odpm_communities/documents/page/odpm_comm_023301.hcsp

[6] Carey, A., Briefing: Bringing brownfield sites back into use, *Municipal Engineer*, **157 (ME4)**, 231-237, 2004.

[7] CPRE [Council for the Protection of Rural England], Housing supply. http://www.cpre.org.uk/campaigns/planning/housing-supply/index.htm

[8] GP [Green Places], CPRE warns of loss of green belt, *Green Places*, **16(6)**, 2005.

[9] Grimski, D. & Ferber, U., Urban brownfields in Europe, *Land contamination and reclamation*, **9(1)**, 143-148, 2001.

[10] OECD [Organisation for Economic Co-operation and Development], *Urban brownfields*, OECD: Paris, 2000.

[11] Moore, N. M., From indigenous industry to foreign finance: The changing face of Dublin Docklands, *Land Use Policy*, **19**, 325-331, 2002.

[12] Gatrell, J. D. & Jenson, R. R., Growth through greening: Developing and assessing alternative economic development programmes, *Applied geography*, **22**, 331-350, 2002.

[13] Harrison, C. & Davis, G., Conserving biodiversity that matters: Practitioners perspectives on brownfield development and urban nature conservation in London, *Journal of environmental management*, **65**, 95-108, 2002.

[14] Nijkamp, P., Rodenburg, C. A. & Wagentonk, A. J., Success factors for sustainable urban brownfield development: A comparative approach to polluted sites, *Ecological economics*, **40**, 235-252, 2002.

[15] Layne, J., Marked for success??? The Winnipeg Core Area Initiative's approach to urban regeneration, *Canadian Journal of Regional Science*, **Summer**, 249-278, 2000.

[16] DLEUC [Department of Land Economy in the University of Cambridge], *Evaluation of the single regeneration budget challenge fund: An examination of baseline issues*, A report for the DETR, DLE: Cambridge, 1999.

[17] SGN [Smart Growth Newtork], Fairfield ecological park, Baltimore, Maryland. www.smartgrowth.org/casestudies/ecoin_fairfield.html

[18] Amekudzi, A. & Fomunung, I., Integrating brownfields redevelopment with transportation planning, *Journal of Urban Planning and Development*, **130(4)**, 204-212, 2004.

[19] Burnham-Howard, C. E., Building on brownfields: Predicted effects of new liability protections for prospective purchasers and an exploration of other redevelopment incentives, *Journal of Professional issues in engineering education and practice*, **130(3)**, 212-225, 2004.

[20] WHCEQ [White House Council on Environmental Quality], Over 1,000 to be trained for environmental jobs in brownfields communities nationwide, *CEQ E-Notes*, **2(13)**, 2004.

[21] English, M., A guide for smart growth. *Forum for applied research and public policy*, **14(3)**, 35-39, 1999.

[22] DETR [Department of the Environment, Transport & the Regions], *Derelict land surveys in 1988 and 1993*, HMSO: London, 1998.

[23] De Sousa, C. A. Turning brownfields into green space in the city of Toronto, *Landscape and urban planning*, **62**, 181-198, 2003.

[24] Jensen, M. B., Persson, B., Guldager, S., Reeh, U. & Nilsson, K., Green structure and sustainability: Developing a tool for local planning. *Landscape and urban planning*, **52**, 117-133, 2000.

[25] Dunnett, N., Swanwich, C. & Woolley, H., *Improving urban parks, play areas and green spaces*, HMSO: London, 2002.

[26] ODPM [Office of the Deputy Prime Minister], *Creating sustainable communities: Greening the gateway*, HMSO: London, 2004.

[27] Brunsen, M. W. & Shelby, B., Assessing recreational and scenic quality. How does new forestry rate? *Journal of Forestry*, **90**, 37-41, 1992.

[28] Akbar, K. F., Hale, W. H. G. & Headley, A. D., Assessment of scenic beauty of the roadside vegetation in northern England, *Landscape and Urban Planning*, **63**, 139-144, 2003.

[29] Lang, D. A. & McNeil, S., Clean it up and will they come? Defining successful brownfield development, *Journal of urban planning and development*, **130(2)**, 101-108, 2004.

[30] Anderson, L. & Cordell, H., Residential property values improving with trees. *Scandinavian Journal of Applied Forestry*, **9**, 162-166, 1985.

[31] Morancho, A. B., A hedonistic valuation of urban green areas, *Landscape and urban planning*, **66**, 35-41, 2003.

[32] Holden, R., Park and pride: Creating an urban park., *The architects journal*, **July**, 2001.

[33] Tyrvainen, L. & Miettinen, A., Property pries and urban forest amenities, *Journal of environmental economic and management*, **39**, 205-223, 2000.

[34] Zetter, J., Frameworks for sound urban policy implementation, *Town & Country Planning*, **74** (2), 57-60, 2005.

[35] Howland, M. What makes for a successful brownfield redevelopment? Three Baltimore case studies. www.smartgrowth.umd.edu/research/pdf/HowlandBrowns.

[36] Lang, D. A. & McNeil, S., Brownfield development: Tools for stewardship, *Journal of urban planning and development*, **130(2)**, 109-116, 2004.

[37] UTF [Urban Task Force], *Towards an urban renaissance*, E. & F. N. Spoon: London, 1999.

[38] Andrew, A. S., Brownfield redevelopment: A State led reform of superfund liability. *Natural Resources and Environment*, **10(3)**, 27-31, 1996.

[39] Strother, P. C., Brownfields of dreams in the old dominion: Redeveloping brownfields in Virginia, *William and Mary Environmental Law and Policy Review*, **23**, 269-304, 2000.

[40] EP [English Partnerships], *Mixing places: A guide to good practice in undertaking mixed development schemes*, Urban villages forum, English Partnerships: London, 1999.

[41] CLARINET [Contaminated Land Rehabilitation Network for Environmental Technologies], The report of the CLARINET working group on brownfield redevelopment 1998 – 2001. www.clarinet.at

[42] Alker, S. & McDonald, A., Incorporating sustainable development into redevelopment, *Sustainable Development*, **11**, 171-182, 2003.

[43] Buttel, F., Ecological modernisation as a social theory, *Gerforum*, **31**, 57-65, 2000.

[44] Wolman, H. L., Ford III, C. C. & Hill, E., Evaluating the success of urban success stories, *Urban Studies*, **31**, 835-850, 1994.

[45] Wernstedt, K., Heberle, L., Alberini, A. & Meyer, P., The brownfields phenomenon: Much ado about something or the taming of the shrewd? Discussion paper 04-46. www.rff.org

[46] Stenner, R. D., Hull, R. N. & Willes, R. F. Involving stakeholders to achieve successful development of brownfield sites. *Brownfield sites: Assessment, rehabilitation and development* eds. C. A. Brebbia, D. Almorza, & H. Klapperich, WIT Press: Southampton, 2002.

[47] Villavaso, S., Sinel, P. & Dauterive, L. The brownfields cookbook: A redevelopment guide. *Brownfield sites: Assessment, rehabilitation and development* eds. C. A. Brebbia, D. Almorza, & H. Klapperich, WIT Press: Southampton, 2002.

[48] Agrawal, A., Common property institutions and sustainable governance of resources, *World Development*, **29(10)**, 1649-1672, 2001.

[49] Pagdee, A., Kim, Y. & Daugherty, P. J., What makes community forest management successful: A meta-study from community forests throughout the world, *Society and Natural Resources*, **19(1)**, 33-52, 2006.

Evaluating brownfield redevelopment projects: a review of existing sustainability indicator tools and their adoption by the UK development industry

K. Pediaditi, W. Wehrmeyer & K. Burningham
Centre for Environmental Strategy, University of Surrey, Guildford, UK

Abstract

There is a plethora of sustainability indicator tools, yet their capacity to monitor the sustainability of Brownfield redevelopment projects is questionable. Furthermore, there is a lack of knowledge about the actual extent to which these tools are being utilised by the development industry. This paper reviews the applicability of sustainability indicator tools to Brownfield redevelopment projects and assesses their capacity to evaluate sustainability throughout the life cycle of such projects, from project conception, design and planning, through construction and remediation to operation and recycling. The results of a survey sent to more than 900 developers in the UK exploring how, and to what extent they assess or monitor the sustainability of their developments are presented. Results indicate that currently there is no sustainability indicator tool designed specifically for brownfield redevelopment projects, and that uptake of existing tools in the development industry is poor. The paper concludes by considering developers' own proposals to overcome the barriers to the adoption of such tools.

Keywords: sustainability, indicators, monitoring, assessment.

1 Introduction

A recent study by the SUE-MoT Consortium identified 632 tools for the assessment of urban sustainability [1]. Innes and Booher [2, pg174] state, referring to the sustainability evaluation tool development, *'this movement is developing so quickly that little has as yet been published documenting, much less critically evaluating, these experiments or assessing their impact. The*

WIT Transactions on Ecology and the Environment, Vol 94, © 2006 WIT Press
www.witpress.com, ISSN 1743-3541 (on-line)
doi:10.2495/BF060061

internet is a much better source than the library for finding out about much of this work, although its descriptions are sketchy and reflect the image each group want to offer'. Mitchell [3] comments on the ad hoc development of tools and sustainability indicators, whereas [4] an overlap between tools. Both [1] and [5] point out the lack of knowledge with regard to the extent of use of sustainability evaluation tools.

Pediaditi *et al.* [6] define the essential characteristics of a BRP sustainability evaluation tool as being:

- **Holistic**: Assessing environmental, social and economic aspects of a BRP.
- **Site and Project Specific**: Assessing the sustainability of a BRP at the wider development scale rather than only focusing purely on the building structure as well as being able to take into account site conditions.
- **Long term**: Assessing the sustainability of a BRP throughout its land use life cycle, including the planning and design phase, the construction and remediation phase as well as operation phase.
- **Participatory**: Enabling evaluation users to make their values and risk perceptions explicit as well as to develop their own sustainability indicators based on those. (Transparency being an essential element of such a process).
- **Integrated within existing decision making processes**: For example the UK land use planning process and developer project management.

Therefore 27 existing sustainability evaluation tools, (which are representative of the different types of tools; See References 7 to 32) were reviewed with regard to their relevance to BRP based on the above criteria, and their extent to which they are used by the UK development industry. The paper concludes with an outline of the perceived barriers to BRP sustainability evaluation and developers' own recommendations for enabling the wider adoption of such practice.

2 Methodology

In order to conduct the tool evaluation, a secondary review of the literature as well as examination of the information available on the tools websites was used, and juxtaposed against the aforementioned criteria. The majority of these tools do not disclose the actual indicators or benchmarks used therefore a detailed evaluation was not possible. Where relevant the authors' conclusions are supplemented with that of other indicator review research.

In order to establish the extent of long term BRP sustainability monitoring a questionnaire was sent out to 987 UK developers both commercial and house-builders. A 9.5% usable response rate was achieved. (Responses were more representative of the medium to large scale developers, thus the sample is more representative in terms of annual volume of development.) Developers were questioned whether they carry out long term sustainability monitoring (i.e., from project inception to development operation) and which methods they use.

Follow up semi structured personal interviews were carried out with 10 developers to establish what incentives and barriers they perceived to the adoption of sustainability evaluation practices, as well as to obtain recommendations on measures required to enhance the wider use of such tools.

3 Review of existing sustainability evaluation tools

3.1 Holistic approach

Pediaditi *et al.* [6] and Hardi and Zdan [33] emphasise the need to adopt a holistic approach when assessing sustainability by giving equal consideration to social, environmental and economic issues. However, when examining Table 1 it is apparent that there is a predominance of solely environmental evaluation tools, which are thus inappropriate for the evaluation of BRP. This conclusion is also drawn by [34] who attribute the lack of emphasis on social indicators to the lack of consensus surrounding what significant social impacts are. A particular gap evident in Table 1 (and confirmed by [35]) is the development of social indicators with regard to building assessment methods.

3.2 Site or BRP specific.

An overview of different tools identifies that there is diversity with regard to the different scales they address, ranging from building component to neighbourhood or city scale, (classified as wider community in Table 1). Cooper [35] assert that the spatial dimension plays an important role and can hinder the integration between different tool methodologies. Pediaditi *et al.* [6] establishes that the appropriate scale to evaluate the sustainability of a BRP is (as classified in Table 1) the development scale, which encompasses the assessment of the developments wider impacts on a locality rather than focusing entirely on the impact of the buildings itself. The majority of reviewed tools focused on evaluating the environmental performance of buildings and at a more detailed level, material and components, mainly based on Life Cycle Assessment methodologies. Therefore, based on the above it is concluded that out of the 27 tools reviewed only 3 tools [13, 15, 28] are potentially relevant for the evaluation of BRP.

3.3 Potential for sustainability evaluation throughout land use life-cycle.

As pointed out in [36] and [4] there are very different tools and assessment approaches in planning (strategic and local) and between the different sectors of development projects i.e. design, construction, operation and all they entail. What is apparent from Table 1 (and supported by [37]) is that there is a tendency for the tools to focus on the initial planning and design phases, which overshadow the sustainability assessment needs of the construction and operational phases of a development. Even tools which claim relevance to the operation phase mainly consist of one off assessments rather than ongoing monitoring schemes. However, it was established in [6] that there is a need to evaluate the different impacts of a BRP throughout its life cycle and thus the only potentially relevant tools are [13, 15, 28].

BRPs often have the characteristic of requiring remediation which may also have sustainability impacts [38] and thus should be evaluated. However, a review of the tools in Table 1 as well as the literature [39–42], it was established that there is no tool designed to evaluate holistically the sustainability of remediation and reclamation schemes. Although theoretical frameworks are available for remediation option selection, (for example [41–45]), they have limited practical application as decisions are currently based on [46] model procedures for the management of land contamination which do not take sustainability into consideration. Furthermore, these frameworks do not assess the sustainability of different options, neither the sustainability impact of remediation processes particular to site circumstances or post BRP completion. (Apart from 45 which is a checklist for the assessment of BRP funding applications which include assessment criteria of remediation processes.) There is a clear gap in the availability of sustainability indicators for remediation and reclamation processes, which deserves further research.

3.4 Participation (stakeholder involvement in BRP sustainability evaluation)

Defining and evaluating sustainability involves value based decisions, and thus participation methods are required to enable stakeholders to define the criteria with which to carry out BRP evaluations [5, 6, 47]. However the review of the tools in Table 1, established that none of the existing tools relevant to the development scale adopted this approach. This is in accordance with [5] who claim that traditional approaches to sustainability indicator development are characterised by rigid scientific method which in turn reflects the reductionist mind set of evaluation developers. Brandon *et al.* [48] referring specifically to built environment sustainability assessments express the limitations of reductionist science to capture the interactions between the parts of systems and feedback which make the resolution of insoluble trains difficult or impossible.

One of the criteria of 'good' participatory decision making as well as risk communication is transparency [49]. From the author's review all the tools relevant at the development scale apart from [28] are developed by consultancies or patented and thus required a fee to be carried out. As a result few of these tools' assessment criteria or benchmarks are disclosed resulting in a loss of transparency, making them less appropriate for the evaluation of BRP.

3.5 Evaluation tool integration with existing planning and BRP decision making processes.

It is important to integrate sustainability evaluation processes within existing planning processes to enable the results of the evaluation to impinge on the nature of the development [6]. However, the review of tools in Table 1 it was identified that only tools [15, 28, 31, 32] made any reference to UK planning policy or Regulations.

Table 1: Classification of existing urban sustainability tools according to their scale, land use life-cycle phase and holistic approach.

	Planning and Design Phase			Construction Phase			Operation Phase		
	Environmental	Social	Economic	Environmental	Social	Economic	Environmental	Social	Economic
Materials/ Components Scale	8,9,14,20,22,23 30			7, 8, 20			8, 20		
Building Scale	8, 10, 11, 13, 14, 17, 20, 21, 22, 24, 25, 28, 30, 31, 32	13, 28	13, 28	7, 13, 18, 20, 21, 24, 25, 28	13, 28	13, 18, 28	8, 13, 18, 21, 24, 25, 28, 30	13, 20, 28	13, 18, 28
Development Scale	***13***, ***15***, **24**, ***28***, **31, 32**	***13***, ***15***, ***28*** **32**	***13***, ***15***, ***28***	**7**, ***13***, ***15***, ***28***	**7**, ***13***, ***15***, ***28***	***13***, ***15***, ***28***	***13***, ***15***, **19**, ***28***	***13***, ***15***, ***28***	***13***, ***15***, ***28***
Wider Community Scale	12, 16, 26, 29	12, 26, 27, 29	12, 16, 26, 27, 29	12	12	12	12, 26	12, 26	12, 26

The most relevant evaluation tools are in both **bold** and ***italics***. They are tools which are applicable to the development scale, look at issues holistically i.e., have environmental, social and economic indicators, and evaluate the sustainability throughout a BRP life cycle phase (planning, construction and operation). Therefore, the only potentially relevant tools as [13, 15, 28]

All numbers relate to references.

Many numbers in a cell indicate potential overlap of tools, whereas a gap in a cell identifies areas for further research.

It can be concluded that the only partially appropriate existing evaluation tools reviewed are [13, 15, 28], with [28] being more suitable due to its transparency, with disclosed and freely available criteria as well as UK planning policy relevance. However, this review is not exhaustive and therefore it was considered important to carry out a UK National Developers Survey to establish to whether sustainability evaluation of BRP was being carried out and which tools were used.

4 Sustainability evaluation practice of the UK development industry

Figure 1 illustrates that half the developers have never carried out long term sustainability monitoring whereas only 17% always do. The survey included an open ended question which asked developers to specify the type of sustainability monitoring which they used. The responses were unclear, many developers did not specify, or provided responses ranging from cost control to the use of independent consultants. Follow up interviews with 10 developers established that they conduct a lot of monitoring regarding output deliverables, cost control, market research etc as part of project management. However, what was also clear from their answers was a lack of understanding of the meaning of long term sustainability monitoring, for example:

> *'we monitor all our operations closely, looking at cost control, return on investment, quality control, safety audits I could spend all day listing the monitoring we do'.*

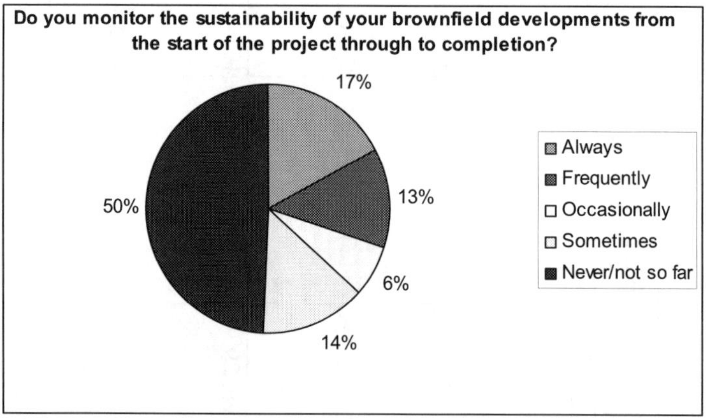

Figure 1: Percentage of developers claiming to carry out long term sustainability monitoring.

The above results demonstrate that there is little sustainability monitoring going on in the development industry. The lack of understanding of the phrase potentially indicates that even the low percentage of developers who claimed to be undertaking sustainability monitoring may in fact be referring to other types of monitoring not relevant to the tools reviewed in Section 3. This confirms the fact that despite the wide range of available sustainability tools there is a lack of uptake and use by the development industry. Therefore, in the interviews developers were questioned with regard to the drivers and barriers to sustainability evaluation practice and were asked to make recommendations for evaluation tool developers and policy makers.

5 Drivers, barriers and recommendations for the wider use of sustainability evaluation tools

Nine out of the 10 developers interviewed stated they saw themselves undertaking sustainability monitoring within the next few years, some drawing parallels between the status of sustainability monitoring today and health and safety monitoring ten years ago. All developers stated policy and changing regulations as being the major driver and examples were provided with regard to the new more demanding energy efficiency regulations. Market pressure was seen as a driver by three developers but was classified as less important in relation to changing regulations.

The perceived barriers to the adoption of sustainability evaluation practices have been summarised in Table 2. All developers commented on the lack of a level playing field in development decision making and emphasised the need for a structured process to enable assessment of development planning applications. Developers, pointed out that the only incentives for carrying out voluntary evaluations would be their potential to smooth the planning process or to save time and resources. A main recommendation by all developers was that any sustainability evaluation tool should be compatible with project time lines and be context specific and simple to overcome the procedural barriers.

Table 2: Barriers to sustainability evaluation.

Barriers to adoption	Procedural limitations	Tool limitations
▪ Lack of understanding of sustainability ▪ Lack of market demand ▪ Lack of enforcement/ resources & skills ▪ Too many tools resulting in lack of confidence in them ▪ Build & Forget development culture	▪ Lack of time ▪ Lack of a structured process to follow ▪ Lack of communication ▪ Lack of ownership of the assessment process ▪ Lack of integration of existing tools with planning processes e.g., planning application process, EIA, SEA, SA	▪ Scope of assessments limited to building performance ▪ Scope of assessments mostly covering environmental issues ▪ Lack of context specific assessments ▪ Lack of measurable benchmarks ▪ Output approach to monitoring

Overall, developers expressed confusion with regard to which tools to use, and that instead of more indicators, a process which would integrate existing tools into the planning and development decision making processes was needed. Importantly developers argued that they would not hesitate to fund sustainability assessments provided they were mandatory for all developers and that they helped bring structure to existing planning decision making.

6 Conclusions

In conclusion the review of existing tools identified that there are no evaluation methods directly relevant to assess the long tem sustainability of BRP in a context specific, holistic and participatory way. A particular gap was found with regard to the availability of methods able to assess the sustainability of remediation processes linked to development decision making. The national developers survey indicated that uptake of sustainability evaluation was minimal, with confusion apparent over the meaning of sustainability evaluation. On the positive side, developers did see themselves carrying out such processes in the short term future and made recommendations to overcome the procedural issues which have hindered uptake thus far. In response to these results the authors have developed and trailed the Redevelopment Assessment Framework RAF described in [6] and [38]. This is a process which is designed to integrate existing evaluation tools into the UK planning decision making processes in a participatory way and which aims to develop context specific and holistic assessments of BRP with the aim of increasing industry utilisation.

Acknowledgements

This research has been carried out as part of the Sustainable Urban Regeneration Integrated Management (SUBR:IM) Research consortium which is funded by the Engineering and Physical Sciences Research Council EPSRC-GR/S18809/01. The authors would like to thank Tim Dixon and the University of Reading for their collaboration with the National Developers Survey.

References

[1] SUE MoT, 2004, http://www.sue-mot.org.uk

[2] Innes, J. & Booher, D., Indicators for sustainable communities: A strategy building on complexity and distributed intelligence, *Planning Theory & Practice*, **1 (2),** pp173-186, 2000.

[3] Mitchell, G., Problems and fundamentals of sustainable development indicators, *Sustainable Development*, **4**, pp 1-11, 1996.

[4] Deaking, M., Huovila, P., Rao, S., Sunikka, M., & Vreeker, R., The assessment of sustainable urban development, *Building Research and Information*, **30 (2)**, pp 95-108, 2002.

[5] Bell, S. & Morse, S., *Sustainability Indicators, Measuring the Immeasurable*, Earthscan, London, 1999.

[6] Pediaditi, K., Wehrmeyer, W. & Chenoweth, J., Sustainability Evaluation for brownfield redevelopment, *Engineering Sustainability*, **159 (ES1)**, pp 3-10, March 2006.

[7] CEEQUAL, Civil Engineering Environmental Quality Award www.ceequal.com

[8] BRE- Methodology for environmental profile of Construction materials, components and Buildings www.bre.co.uk
[9] The Boustead Model www.boustead-consulting.co.uk
[10] Building Design Advisor http://gaia.lbl.gov/bda
[11] DOE-2.2, http://simulationresearch.lbl.gov
[12] Community Sustainability Assessment, http://gen.ecovillage.org/activities/csa/English/toc.php
[13] SPeAR The Sustainable Project Appraisal Routine http://www.arup.com/environment/feature.cfm?pageid=1685
[14] Athena www.athenaSMI.ca
[15] BRE sustainability checklist for developments www.bre.co.uk
[16] City Green www.americanforests.org
[17] ECOTECT www.spu1.com
[18] Ecopro http://research.scpm.salford.ac.uk/resources/docs/DOCUMENTS/BQMasterlist2rev.xls
[19] Eco Cal www.bestfootforward.com/ecocal.htm
[20] ENVEST 2 www.bre.co.uk
[21] The Hong Kong Building Environmental Assessment Method HK-BEAM http://www.bse.polyu.edu.hk/Research_Centre/BEp/hkbeam/HK
[22] Green Building Challenge www.greenbuilding.ca
[23] GaBi 4 www.pe-europe.com
[24] LEED Leadership in Energy and Environmental Design www.usgbc.org
[25] Minnesota Sustainable Design Guide MSDG http://www.develop.csbr.umn.edu/msdg2/MSDG/guide2.html
[26] Planning for community Energy, Economic and Environmental Sustainability PLACES www.energy.ca.gov/places/
[27] Social Costs of Alternative Land Development Scenarios www.fhwa.dot.gov/scalds/scalds.html
[28] SEEDA sustainability Checklist www.sustainability-checklist.co.uk
[29] SPARTACUS System for Planning and Research in Towns and Cities for Urban Sustainability www.raumplanung.uni-dormund.de/irpud/pro/sparta/sparta.htm
[30] TEAN, Toolfor environmental Analysis and Management www.ecobilan.com
[31] BREEAM Building Research Establishment Environmental Assessment Method www.breea.org
[32] EcoHomes www.bre.co.uk
[33] Hardi, P. & Zdan, T., *Assessing Sustainable Development: Principles in Practice*, IISD, Canada, 1997.
[34] Levett & Therivel, *Report to the SUE MoT consortium: Sustainable Urban Environment- Metrics, Models and Toolkits: Analysis of sustainability/social tools*, Oxford, Unpublished, June 2004.
[35] Cooper, I., Environmental assessment methods for use at the building and city scales: constructing bridges or identifying common ground? In Brandon, P., Lombardi, P. & Bentivenga, V., (eds), *Evaluation in the Built Environment for Sustainability*, E & EF Spon, London, 1997.

[36] Curwell, S., Hamilton, A. & Cooper, I., The BEQUEST framework: towards sustainable urban development, *Building Research and Information*, **26 (1)**, pp56-65, 1998.

[37] Deaking, M., Curwell, S. & Lomberdi, P., Sustainable urban development: the framework and directory of assessment methods, *Journal of Environmental Policy and Management*, **4 (2)**, June, pp171-197, 2002.

[38] Pediaditi, K., Wehrmeyer, W. & Chenoweth, J., Monitoring the sustainability of Brownfield redevelopment Projects. The Redevelopment Assessment Framework (RAF), *Contaminated Land & Reclamation,* **13 (2)**, pg 173-183, 2005.

[39] Bardos, R., Nathanail, C. & Weenk, A., *Assessing the wider environmental value of remediating land contamination, A review. R&D technical Report P238*, Environment Agency, Swindon, 1999.

[40] Bardos, R., Kearney, T., Nathanail, C., Weenk, A. & Martin, I., Assessing the wider environmental value of remediating land contamination, 7th international FZK/TNO, Conference on Contaminated Soil 'ConSoil', http://www.r3environmental.co.uk/CONSOIL%202000%20Lecture%20Session%20B2%20Bardos%20et%20al.PDF

[41] Vik, E., Bardos, P., Brogan, J., Edwards, D., Gondi, F., Henrysson, T., Jensen, B., Jorge, C., Marrioti, C., Nathanial, P. & Papassiopi, N., Towards a Framework of selecting remediation technologies for contaminated sites, *Land Contamination and Reclamation*, **9(1)** pp 119-127, 2001.

[42] Rudland D. & Jackson, S., *Selection of remedial treatments for contaminated land. A guide to good practice*, CIRIA C622, London, 2004.

[43] Harris M., Herbert, S. & Smith, M., *Remedial Treatment for contaminated land. Classification and selection of remedial methods*, VI, SP104, CIRIA, London, 1995.

[44] SAFE GROUNDS, Good practice guidance for the management of contaminated land on nuclear and defence sites, 2002 www.safegrounds.com

[45] RESCUE, Best Practice guidance for sustainable brownfield regeneration, http://www.rescue-urope.com/download/reports/RESCUE%20Manual.pdf

[46] DEFRA & EA, Model procedures for the management of land contamination, Contaminated Land Report 11, 2004, www.environment-agency.gov.uk/commondata/105385/**model_procedures**_881483.pdf

[47] Ukaga, O. & Maser, C., *Evaluating Sustainable Development, Giving People a voice in their destiny*, Stylus Publishing Virginia, 2004.

[48] Brandon, P., Lombardi, P. & Bentivenga, V., (eds) *Evaluation in the built environment for sustainability*, E & EF Spon, London, 1997.

Section 3
Risk assessment and management

"ARA-CoS$_{SSLs}$" risk analysis approach (on a computer worksheet basis) for contaminated sites: implementation and case study application

M. Giavazzi & F. Tatàno
Environmental Science Faculty, University of Urbino "Carlo Bo", Italy

Abstract

Within the increasing attention to absolute risk analysis at Italian level, as a University research study, the definable "ARA-CoS$_{SSLs}$" approach for human health ("Absolute Risk Analysis for Contaminated Sites", on "Soil Screening Levels" basis) has been derived (as an initial, simplified version) and directly implemented on a series of interactive computer worksheets. The basic conditions of the "ARA-CoS$_{SSLs}$" approach derivation are: 1) the predominant reference to a fundamental international scientific protocol, also considering its temporal evolution (the US.EPA "SSLs"); 2) a modular, open, common for users, and easily extensible computer calculation. This paper reports on the basic aspects and the specific calculation structure of the "ARA-CoS$_{SSLs}$" approach; additionally, a simplified case-study application is summarised.
Keywords: absolute risk assessment, case-study, contaminated sites, exposure pathways and scenarios, model implementation.

1 Introduction

Within the integrated management for contaminated site identification, characterisation and final remediation [1], an indispensable step deals with the application of reliable and scientifically based absolute risk analysis criteria [2], in order to: 1) evaluate the contamination level of environmental media at a given suspected site, as quantification of the associated risks for human and/or ecological targets; 2) eventually, define risk-based residual chemical concentrations, if a remedial action is required at the given site. As far as the

WIT Transactions on Ecology and the Environment, Vol 94, © 2006 WIT Press
www.witpress.com, ISSN 1743-3541 (on-line)
doi:10.2495/BF060071

Italian situation is concerned, despite a restricted regulative application of absolute risk analysis up until the present time forced by the existing "D.M. No. 471/99" Ministerial Decree on contaminated soil management [2, 3], some international and/or national software tools for absolute risk analysis are currently adopted or at least recognized in Italy, *inter alia*: "RBCA Tool Kit for Chemical Releases", based on the specific ASTM Standard for "RBCA, Risk-Based Corrective Action" [4]; "ROME", developed by the Italian Environmental Protection Agency [5]; and "GIUDITTA", proposed by the Province of Milan [6]. Additionally, the Italian Environmental Protection Agency and Technical Services has recently produced a technical document on basic methodological criteria for absolute risk analysis applications at Italian contaminated sites [7]. With this increasing attention to the absolute risk analysis at Italian level, the definable "ARA-CoS$_{SSLs}$" modelling approach for human health ("Absolute Risk Analysis for Contaminated Sites", on "Soil Screening Levels" basis) is synthetically described in this paper, as derived (exclusively due to a University research study) in its initial, simplified version and implemented on a series of interactive computer worksheets. Basic conditions for the "ARA-CoS$_{SSls}$" approach derivation have been expressly: 1) the predominant reference to a fundamental and unique international scientific protocol, also considering its temporal evolution (the US.EPA "SSLs, Soil Screening Levels" procedure); 2) a modular, open, common for users, and easily extensible computer calculation structure. A simplified case-study application of the "ARA-CoS$_{SSls}$" modelling approach is also finally summarised.

2 The scientific reference procedure: US.EPA SSLs

The "SSLs" procedure has been defined and temporally updated by US.EPA as a tiered sanitary risk assessment framework for deriving risk-based (generic or site-specific) soil screening levels, according to three, possible site information approaches [10]: a first, generic approach (reasonable, conservative assumptions); a second, simple site-specific methodology; a third, more detailed site-specific modelling approach. The original "US.EPA 1996 SSLs" procedure [8, 9] was expressly focused on residential land use scenario (with on-site residents as receptors), including: 1) the detailed derivation of risk equations related to the assumed pathways of concern (direct soil ingestion, outdoor inhalation of fugitive dusts and volatiles, and ingestion of contaminated ground water); 2) the accurate summary of default values (according to the generic assessment approach) for human exposure factors, soil and ground water characterising parameters, and meteorological data; 3) the database of all relevant contaminant physical-chemical and toxicological parameters; and 4), the definition of sampling and representative concentration estimation procedures for surface (usually top 2 cm) and subsurface soils, with related statistical data quality tests (for surface soil). Then, the supplemental "US.EPA 2002 SSLs" guidance [10] has updated and extended the original procedure, synthetically according to the following, main aspects: 1) as related to the previous residential scenario, new risk equations for the concurrent exposure via ingestion and dermal absorption, a new site-specific modelling for the volatile migration into indoor air, and updated dispersion modelling data for the air exposure model; 2) additional, detailed consideration of a commercial/industrial scenario (outdoor and indoor workers as targets) and a construction scenario (construction workers and off-site residents as receptors); and finally 3), additional statistic option of the "$UCL_{95\%}$, Upper Confidence Limit" for the proper definition of representative chemical concentrations in surface and subsurface soils.

3 "ARA-CoS_{SSLs}" modelling approach

3.1 General calculation structure and contaminants of concern

The "ARA-CoS_{SSLs}" absolute sanitary risk analysis model, with a modular structure based on the US.EPA "SSLs" procedure (see Section 2), has been directly implemented in Microsoft Excel® format [11]. According to the usual international approaches for absolute risk analysis [7], "ARA-CoS_{SSLs}" model performs two risk assessment modes: 1) the "direct" (or "forward") evaluation of health risks associated with site contamination; and 2) the "inverse" (or "backward") calculation of risk-based soil quality limits or cleanup targets. Additionally, for both modes, two assessment tiers are considered: a first, "site-generic" assessment level, expressly based on the US.EPA generic, reasonable conservative exposure scenario assumptions (see Section 2); and a second, "site-specific" assessment level, based instead on simple site-specific characterisation data. According to Figure 1, the calculation structure of "ARA-CoS_{SSLs}" model has been specifically arranged with three correlated "Modules" ("I-O, Input-Output", "$A_{B\text{-}DIR}$, Bi-directional Analysis" and "DB, DataBase"), eventually conceptual "Sub-modules", and finally "iWSs, individual Worksheets". With specific reference to the selection of contaminants of concern (from all detected chemicals, at a given suspected site), two different procedures have been currently implemented in "ARA-CoS_{SSLs}" model [11]: the so identifiable "Toxicity Score" approach (as described in [12]), and alternately the "normalised Comparative Hazard Index" approach (as defined in [13]).

3.2 Module "I-O"

It conceptually consists of three Sub-modules ("Input", "Output", and "Statistical Tests"), which are articulated according to the specific iWSs named in Figure 1.

3.2.1 Sub-module "Input"

The interactive "iWS CSM, Conceptual Site Model" (Figure 2) allows to define: the type of risk analysis (generic or simple site-specific), the land use scenario, and the corresponding, significant exposure pathways. At present [11], the following exposure pathways have been included in "ARA-CoS_{SSLs}" model, under either a residential or a commercial/industrial scenario: direct ingestion and dermal absorption (related to surface soil), outdoor inhalation of fugitive dusts (surface soil) and volatiles (related to subsurface soil), and ingestion of contaminated ground water (also subsurface soil). As a flexible and realistic approach, in "ARA-CoS_{SSLs}" model direct ingestion and dermal absorption can be selected either as unique, combined pathway (according indeed to the original US.EPA "SSLs" procedure: see Section 2), or as two independent pathways.

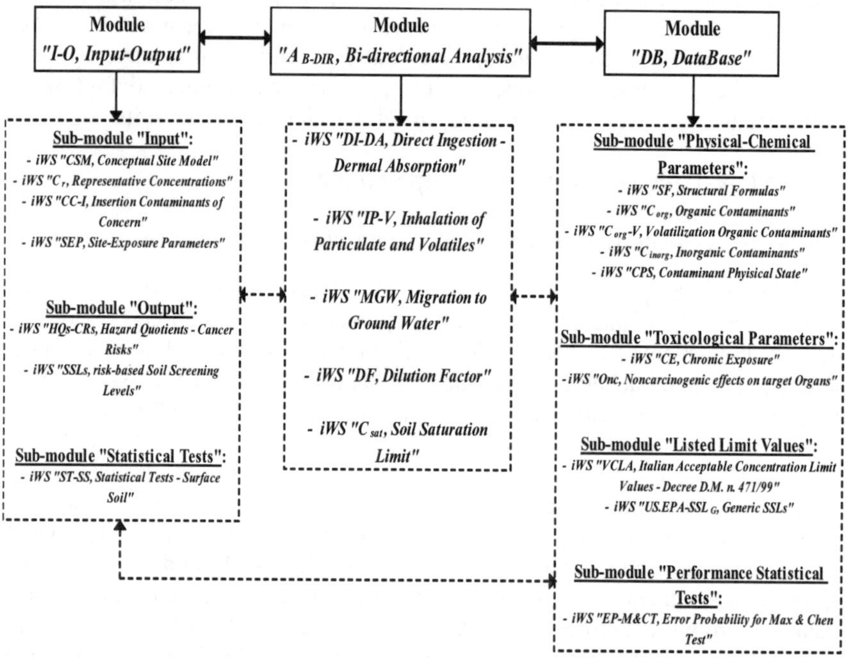

Figure 1: Flow-chart of "ARA-CoS$_{SSLs}$" model.

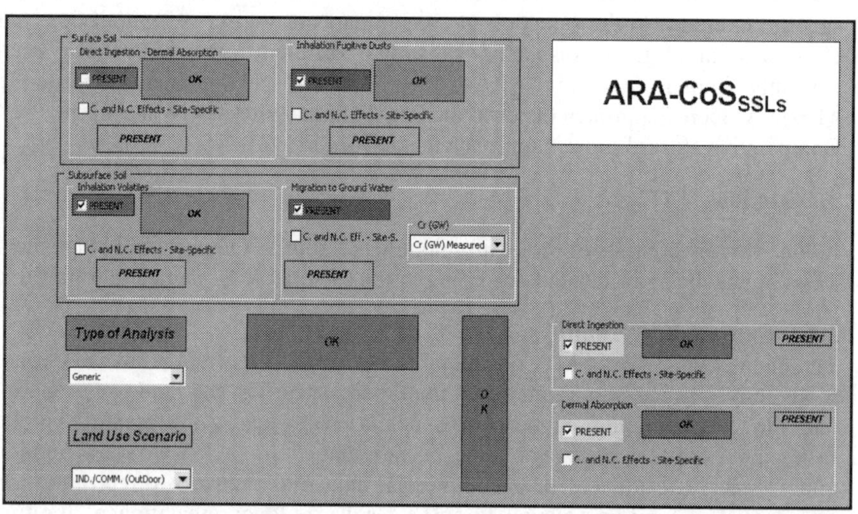

Figure 2: "ARA-CoS$_{SSLs}$" model: view of "iWS CSM, Conceptual Site Model".

Within a "direct" assessment mode, "iWS Cr, Representative Concentrations" provides for the definition of representative chemical concentrations in surface soil (as arithmetic mean or max value, according to a composite or individual sampling procedure [9]), and subsurface soil [11] (as arithmetic mean or max value, according to a proper core length-weighted averaging procedure [9]); additionally, this iWS performs the contaminant indicator selection procedures summarised in Section 3.1. The "iWS CC-I, Insertion Contaminants of Concerns" permits the tabulation of representative concentrations for the selected chemicals of concern, as ppm values for surface and subsurface soil (weight-to-weight ratio, equivalent to mg kg^{-1}) and ground water (weight-to-volume ratio, equivalent to mg l^{-1}). As far as specifically ground water is concerned, "measured" concentrations can be directly inserted ("M" indicative symbol), or alternately "estimated" values ("E" symbol) are automatically generated based on the US.EPA theory of soil/water/air contaminant partition in unsaturated zone and subsequent aquifer dilution [8, 9]. Additionally, this iWS provides for some useful, visual indications [11]: 1) representative soil concentrations above the officially tabulated US.EPA generic "SSLs" [10], and representative ground water concentrations above the officially tabulated Italian generic quality limits [3] (see Section 3.4); and mainly 2), the potential presence of Non-Aqueous Phase Liquid (NAPL) for contaminants (liquid at typical soil temperature) with representative subsurface concentrations exceeding the corresponding "C_{sat}, Soil Saturation Limit" [mg kg^{-1}] [9, 10]. It should be pointed out that, according at least to the theoretical derivation of US.EPA risk approaches related to inhalation of volatiles and migration to ground water [9, 10], in "ARA-CoS_{SSLs}" calculation model chemical soil contents should be adequately inserted as dry weight (with a resulting, conservative risk assessment condition [13]); moreover, again expressly for volatile inhalation and ground water migration pathways [11], subsurface soil chemical concentrations should be theoretically assumed as total, three-phase (soil, water, air) contents [9].

Finally, "iWS SEP, Site-Exposure Parameters" summarises all relevant parameters (exposure, exposed population, soil and aquifer, meteorological conditions) for conducting risk analysis, with default values (under a generic mode analysis) expressly according to US.EPA "SSLs" indications [10].

3.2.2 Sub-module "Output"

It consists of two iWSs (Figure 1). According to the "direct" assessment mode, "iWS HQs-CRs" (Figure 3) calculates the sanitary risks associated with contaminant non-carcinogenic and carcinogenic effects, respectively as "HQ, Hazard Quotient" and "CR, Cancer Risk" [14]. Precisely, risk quantification is carried out either for individual effects (individual toxic effect and pathway of exposure) or for cumulative effects with risk aggregation as follows [8, 9]: 1) for carcinogenic effects, firstly contaminant CRs are simply added for each pathway of exposure, and consequently all pathway CRs are added together; 2) for non-carcinogenic effects, HQs are added only for those contaminants with the same toxic endpoint (human organ). Expressly for inhalation exposures, a visual indication is given in "iWS HQs-CRs" for those chemicals with route-to-route extrapolation of the corresponding toxic parameters ("URF, Unit Risk Factor" [μg^{-1} m^3] and "RfC, Reference Concentration" [mg m^{-3}]) from those related to ingestion (respectively, "SF, Slope Factor" [mg^{-1} kg d] and "RfD, Reference Dose" [mg kg^{-1} d^{-1}]), according to the simple US.EPA derivation approach [9].

According to the "inverse" assessment mode, "iWS SSLs" (Figure 4) defines risk-based Soil Screening Levels (generic or site-specific) for individual pathway and toxic effect (with acceptable individual risk levels assumed as $1 \cdot 10^{-6}$ for CR and 1 for HQ [8, 9, 14]). Expressly for ground water migration pathway, two alternating calculations are possible [11]: 1) the original US.EPA approach [8–10], based on the soil/water/air partition and subsequent dilution theory and the consideration of official acceptable water levels; and 2) a proper "risk-based" approach (see Section 3.3). Finally, also this iWS gives some useful, visual indications [11]: for instance, the eventual, non-reliable condition of resulting calculated SSLs for volatile inhalation and ground water exceeding C_{sat} [9, 10].

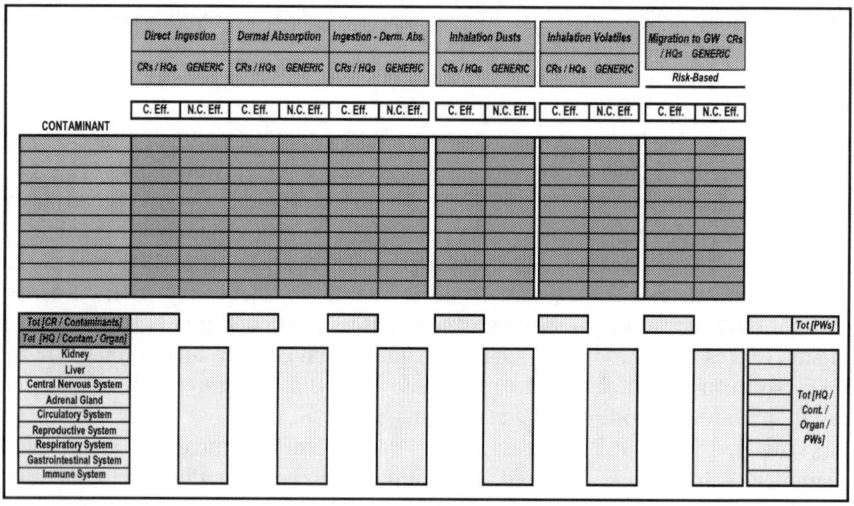

Figure 3: "ARA-CoS$_{SSLs}$" model: partial view of "iWS HQs-CRs". Legend: C., N.C. Eff. = carcinogenic, non-carcinogenic effects; PWs = pathways.

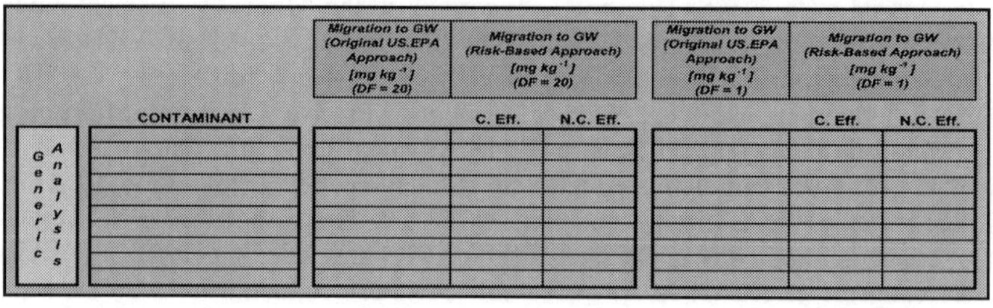

Figure 4: "ARA-CoS$_{SSLs}$" model: partial view of "iWS SSLs". Legend: DF = Dilution Factor.

3.2.3 Sub-module "Statistical tests"

Consisting of a unique iWS (Figure 1), this Sub-module implements the possible execution of "Max" and "Chen" statistical tests for surface soil sampling data validation, within an individual "EA, exposure area" of 0.5 acre (~ 2,000 m^2) [9].

3.3 Module "$A_{B\text{-}DIR}$"

This fundamental Module, properly executing the risk calculation, is composed of the iWSs schematised in Figure 5. Precisely, "iWS DI-DA" implements the US.EPA modelling approaches for direct ingestion and dermal absorption pathways, while "iWS IP-V" concerns with the US.EPA modelling approaches for outdoor inhalation of particulate and volatiles [10]. As far as instead the migration to ground water pathway is concerned, "iWS DF" and "iWS MGW" contain respectively [11]: 1) the US.EPA derivation of "DF, Dilution Factor" [8, 9, 10]; 2) the above mentioned US.EPA soil/water/air partition theory in unsaturated zone [8, 9, 10], in due combination with the US.EPA risk equations related to human intake from contaminated water ingestion [14]. Finally, "iWS C_{sat}" performs the derivation of the soil saturation limit (see Section 3.2.1), with some different parametric default values proper of inhalation of volatiles ("IV" in Figure 6) or migration to ground water ("MGW" in Figure 6) modelling (that is, "θ_w, water-filled soil porosity" and "f_{oc}, fraction organic carbon in soil") [10].

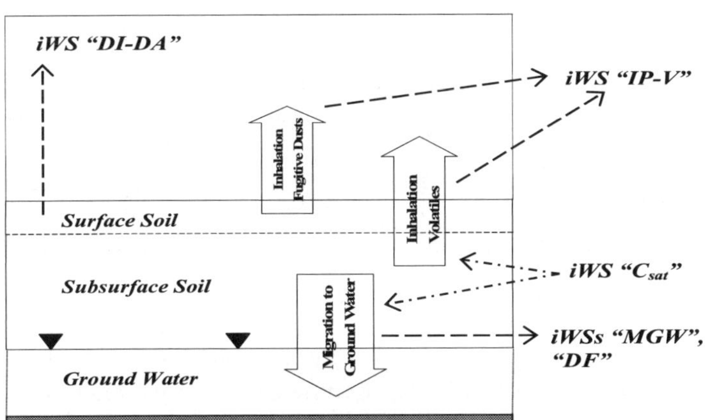

Figure 5: "ARA-CoS_{SSLs}" model: scheme of the environmental sectors of concern for the individual Worksheets within the "$A_{B\text{-}DIR}$" Module.

3.4 Module "DB"

Comprehensive of four conceptual Sub-modules (Figure 1), this Module contains all parametric data necessary for "ARA-CoS_{SSLs}" modelling [11]: contaminant physical-chemical factors (influencing environmental fate and transport) and toxicological parameters, official (Italian and U.S.) soil and ground water generic limits, and error probabilities for "Max" and "Chen" tests (see Section 3.2.3).

4 Summary of a simplified case-study application

A first, simplified application of "ARA-CoS$_{SSLs}$" modelling approach has regarded a service station site with leaching underground storage tanks, and consequential subsoil and ground water detection of BTEX (benzene, toluene, ethylbenzene and xylene), MTBE (methyl ter-butyl ether) and TPH (Total Petroleum Hydrocarbons, C<12 and C>12 fractions). Specific site history and characterisation, developed conceptual site model and consequential assumptions are reported in [15]; expressly, a non-residential scenario has been assumed, with outdoor workers as considered targets. Figures 6 and 7 show the resulting contaminant representative concentrations respectively in soil (surface, subsurface) and ground water. As far as specifically Figure 6 is concerned, it should be pointed out that two "hypothetical" conditions have been considered for subsurface soil [15]: 1) the sampling and analytical condition of desirable preservation, conservation and detection of the "real" three-phase chemical contents (in soil, water and air: "[S-W-A]r" symbol) [9, 16]; 2) the alternating (and probably more realistic) condition of real detection only of the chemical soil and water contents ("[S-W]r" symbol), with consequential, automatic (within "ARA-CoS$_{SSLs}$" model) derivation of the "estimated" total three-phase concentration ("[S-W-A]e" symbol) based on a proper and original mathematical adaptation and extension of the US.EPA contaminant partition theory [9].

	Cr-SS [ppm w/w]	Cr-SSS [ppm w/w - DM] [S-W-A] r *(o [S-W] r)*	*[S-W-A] e / [S-W] r*		*Cr-SSS [ppm w/w - DM] [S-W-A] e*	
			θ_w - f_{oc}			
			IV	*MGW*	*IV*	*MGW*
Chemical Name						
Benzene	1.00E-1	1.00E-1	*1.095E+00*	*1.064E+00*	***1.10E-01***	***1.06E-01***
Ethylbenzene	3.50E+0	3.50E+0	*1.027E+00*	*1.031E+00*	***3.59E+00***	***3.61E+00***
MTBE	8.00E+0	8.00E+0	*1.035E+00*	*1.010E+00*	***8.28E+00***	***8.08E+00***
Toluene	2.60E+0	1.50E+0	*1.043E+00*	*1.043E+00*	***1.56E+00***	***1.56E+00***
TPH (>C8-C10: Aliph.)	8.34E+2	3.73E+2	*1.078E+00*	*1.111E+00*	***4.03E+02***	***4.15E+02***
TPH (>C12-C16: Aliph.)	1.56E+3	7.16E+2	*1.003E+00*	*1.005E+00*	***7.18E+02***	***7.19E+02***
o-Xylene	3.43E+1	3.43E+1	*1.018E+00*	*1.021E+00*	***3.49E+01***	***3.50E+01***

Figure 6: Case-study application of "ARA-CoS$_{SSLs}$" model: partial view of "iWS CC-I" [15]. Legend: SS = surface soil; SSS = subsurface soil; DM = dry weight; [S-W-A]r, [S-W]r, [S-W-A]e= see Section 4; IV, MGW = see Section 3.3 (θ_w, f_{oc} different pathway default values).

	Cr-GW [ppm w/v] - M	Cr-GW [ppm w/v] - E		NAPL Csat [mg/kg]
		[S-W-A] r	*[S-W-A] e*	
Benzene	1.38E+01	1.48E-02	*1.57E-02*	---
Ethylbenzene	2.10E+00	1.83E-01	*1.89E-01*	---
MTBE		1.88E+00	*1.90E+00*	---
Toluene	1.35E+01	1.27E-01	*1.33E-01*	---
TPH (>C8-C10: Aliph.)	8.24E+02	2.65E-01	*2.94E-01*	*8.80E+01*
TPH (>C12-C16: Aliph.)		3.55E-03	*3.57E-03*	*2.29E+01*
o-Xylene	8.60E+00	1.81E+00	*1.85E+00*	---

Figure 7: Case-study application of "ARA-CoS$_{SSLs}$" model: additional, partial view of "iWS CC-I" [15]. Legend: M, E (DF = 20) = see Sect. 3.2.1.

The resulting individual risks, according to a "direct, generic" assessment mode (see Section 3.1), are reported in Table 1. Generic risks associated with direct ingestion and inhalation of fugitive dusts and volatiles appear as negligible, while the risk-based approach for ground water migration shows non-negligible carcinogenic (benzene) and non-carcinogenic (on liver – toluene, on kidney – toluene and ethylbenzene [9], under measured concentrations) generic risks [15]. Finally, risks related to the "estimated" three-phase approach are increased (but not significantly) compared with the hypothetic "real" two-phase approach [15].

Table 1: Simplified case-study application of "ARA-CoSS$_{SLs}$" model: calculated, generic risks (CRs and HQs) [15]. Legend: N/D = not derivable.

	Direct Ingestion		Dermal Absorption		Inhalation of Dusts	
	CR	HQ	CR	HQ	CR	HQ
Benzene	1.73E-09	N/D	N/D	N/D	1.26E-13	N/D
Toluene	N/D	1.14E-05	N/D	N/D	N/D	2.95E-09
Ethylbenzene	N/D	3.08E-05	N/D	N/D	N/D	1.59E-09
o-Xylene	N/D	1.51E-05	N/D	N/D	N/D	2.22E-09
MTBE	N/D	N/D	N/D	N/D	N/D	1.21E-09
TPH (>C8-C10)	N/D	7.34E-03	N/D	N/D	N/D	3.78E-07
TPH (>C12-C16)	N/D	1.37E-02	N/D	N/D	N/D	7.07E-07
	Inhalation of Volatiles ("IV")					
	[S-W-A]r *(or [S-W]r)*		*[S-W-A]e*			
			$\theta_w - f_{oc}$ *(IV modelling)*		$\theta_w - f_{oc}$ *(MGW model.)*	
	CR	HQ	*CR*	*HQ*	*CR*	*HQ*
Benzene	7.05E-08	N/D	*7.72E-08*	*N/D*	*7.50E-08*	*N/D*
Toluene	N/D	6.51E-04	*N/D*	*6.79E-04*	*N/D*	*6.79E-04*
Ethylbenzene	N/D	4.48E-04	*N/D*	*4.60E-04*	*N/D*	*4.62E-04*
o-Xylene	N/D	5.51E-04	*N/D*	*5.61E-04*	*N/D*	*5.63E-04*
MTBE	N/D	4.51E-04	*N/D*	*4.67E-04*	*N/D*	*4.56E-04*
TPH (>C8-C10)	N/D	NAPL	*N/D*	*NAPL*	*N/D*	*NAPL*
TPH (>C12-C16)	N/D	NAPL	*N/D*	*NAPL*	*N/D*	*NAPL*
	Migration to Ground Water ("MGW") (risk-based approach)					
	Measured Cr		Estimated Cr			
			[S-W-A]r *(or [S-W]r)*		*[S-W-A]e*	
	CR	HQ	CR	HQ	*CR*	*HQ*
Benzene	4.77E-03	N/D	5.12E-06	N/D	*5.44E-06*	*N/D*
Toluene	N/D	1.19E+00	N/D	1.12E-02	*N/D*	*1.17E-02*
Ethylbenzene	N/D	3.70E-01	N/D	3.23E-02	*N/D*	*3.33E-02*
o-Xylene	N/D	7.57E-02	N/D	1.60E-02	*N/D*	*1.63E-02*
MTBE	N/D	N/D	N/D	N/D	*N/D*	*N/D*
TPH (>C8-C10)	N/D	NAPL	N/D	NAPL	*N/D*	*NAPL*
TPH (>C12-C16)	N/D	NAPL	N/D	NAPL	*N/D*	*NAPL*

5 Conclusions

As a University research study, the "ARA-CoS$_{SSLs}$" modelling approach for sanitary risk analysis has been derived (as initial version) with predominant reference to the international US.EPA "SSLs" scientific procedure, and directly implemented on a computer worksheet basis. A first, simplified case-study application has been performed. A further "ARA-CoS$_{SSLs}$" model application at a relevant Italian national priority contaminated site is now being carried out.

References

[1] Andreottola, G. & Tatàno, F., Current status and perspectives in contaminated soil management – Keynote Lecture, *CD-ROM Proc. of the "ISWA, International Solid Waste Management" 25th Anniversary World Congress on Waste Management,* UTEC Wien GmbH, Wien, 1995.

[2] D'Aprile, L., Musmeci, L. & Tatàno, F., Soil and groundwater quality criteria for contaminated sites: from the Italian experience to the European situation and perspective, *CD-ROM Proc. of the "European Water Day" Conference,* "EWA, European Water Association" & "ASTEE, French Association Scientific Technical Water Environment", Paris, June, 2005.

[3] Italian Ministry of Environment, *Technical Regulation D.M. No. 471/99 on containment, remediation and environmental restoration of contaminated soils,* Annex of the Italian Official Gazette No. 293, 15.12.1999 (in Italian).

[4] American Society for Testing and Materials (ASTM), *Standard Guide for Risk-Based Corrective Action,* E 2081-00, ASTM International, 2004.

[5] Italian Environmental Protection Agency (ANPA), *ROME: ReasOnable Maximum Exposure,* Version 2.1, Operating Manual, Rome, 2002.

[6] Province of Milan & URS Dames & Moore, *GIUDITTA*, Version 3.0, User's Handbook, Milan, 2002.

[7] Italian Environmental Protection Agency (APAT), *Methodological Criteria for Absolute Risk Analysis Application at Contaminated Sites,* eds. R. Baciocchi, L. D'Aprile & G. Marella, Rome, 2005 (in Italian).

[8] United States Environmental Protection Agency (US.EPA), *Soil Screening Guidance: User's Guide,* Publication 9355.4-23, 1996.

[9] US.EPA, *Soil Screening Guidance: Technical Background Document,* Office of Emergency and Remedial Response, EPA/540/R95/128, 1996.

[10] US.EPA, *Supplemental Guidance for Developing Soil Screening Levels for Superfund Sites,* OSWER 9355.4-24, 2002.

[11] Giavazzi, M. & Tatàno, F., ARA-SiC$_{SSLs}$ absolute risk analysis model for contaminated sites. *RS – Rifiuti Solidi,* XIX(3), pp.165-179, 2005 (in Italian).

[12] LaGrega, M.D., Buckingham, P.L. & Evans, J.C., *Hazardous Waste Management,* Second Edition, McGraw-Hill, USA, 2001.

[13] Zakharova, T., Tatàno, F. & Menshikov, V., Health cancer risk assessment for arsenic exposure in potentially contaminated areas by fertilizer plants: a possible regulatory approach applied to a case study in Moscow Region-Russia. *Regulatory Toxicology and Pharmacology,* **36**, pp. 22-33, 2002.

[14] US.EPA, *Risk Assessment Guidance for Superfund: Vol. I – Human Health Evaluation Manual (Part B),* EPA/540/R-92/003, 9285.7-01B, 1991.

[15] Giavazzi, M., Tatàno, F. & Baronciani, M., ARA-SiC$_{SSLs}$ absolute risk analysis model for contaminated sites: implementation and case-study application. *Groundwater Geoengineering,* **21**, 2006 (in Italian, in press).

[16] Cal-EPA, *Guidance document for the implementation of US.EPA Method 5035: Methodologies for collection, preservation, storage and preparation of soils to be analysed for volatile organic compounds.* Department of Toxic Substances Control, DTSC/Cal-EPA, State of California, November, 2004.

Comparing metal uptake by vegetation under different experimental conditions

D. Sinnett, R. van Herwijnen, C. Kilbride & T. R. Hutchings
Land Regeneration and Urban Greening Research Group, Forest Research, UK

Abstract

Modern planning policies recognise that the provision of green space such as parkland and community woodland offer multifunctional environmental and social benefits which can contribute strongly to the sustainability of towns and cities. A challenge to such development is that the soils of many former industrial sites often contain elevated concentrations of contaminants including potentially toxic elements, for example metals. The bioavailability of metals can be reduced in the contaminated soil prior to greenspace establishment, thus reducing their risk to ecosystems and human health. For the establishment of a safe greenspace in the perspective of the contaminant receptor pathway linkage an understanding of the uptake, compartmentation and toxicological effects of these contaminants to vegetation is important.

We have collated data from hydroponic and contaminated soil pot trials and from field trials on contaminated sites. The data have been used to produce correlations between heavy metal availability, plant uptake and compartmentation of the heavy metals for poplar Fritzi Pauley. From the correlations, models for plant uptake are provided with an assessment on the impact of improved soil conditions and consequent improved plant performance and the metal uptake by these plants. This information may be used to assess the risk for receptors after potential greenspace establishment on heavy metal contaminated soil.

Keywords: Cd, Zn, poplar, metal compartmentation, plant uptake.

WIT Transactions on Ecology and the Environment, Vol 94, © 2006 WIT Press
www.witpress.com, ISSN 1743-3541 (on-line)
doi:10.2495/BF060081

1 Introduction

Greenspace creation has become an increasingly attractive option for the development of former industrial sites. Many sites are found in areas that may have suffered economic decline in recent decades, thus reducing the likelihood of inward investment to deal with the legacy of industrialisation that has taken its toll on the landscape. In addition, these sites are often in densely populated areas that may have little or no provision for amenity land. Their greening can offer a relatively low cost solution for improving the aesthetic appearance and amenity resource of a region and may contribute to encouraging investment and economic growth.

One problem facing developers working in this area is that many brownfield sites contain elevated levels of contaminants including metals such as cadmium, copper, lead and zinc. These metals may originate from a range of sources including smelting, manufacturing and waste disposal operations.

Current legislation in many countries dictates that if contaminants in soils threaten to affect water quality, human or ecological health or property, remedial action must be taken. Threshold concentrations are often set which are based on soil total metal concentrations that may bear little relevance to the likelihood of any threat to the receptor of concern. It is notoriously difficult to predict the toxicological effect of a metal to vegetation, or its uptake into the plant biomass, by its total concentration in soil alone [1]. Often different species may exhibit different tolerances to the same metal or the same species may react differently under different soil conditions.

Cd and Zn represent important metals for greenspace development as they often pose different threats to the vegetation. They are metals with similar geochemical properties and are found together due to their geological associations. Cd is not necessary for plant growth and is often toxic to fauna at concentrations significantly lower than for vegetation. It is also mobile in soils and within the plant and may accumulate in vegetation without significantly affecting plant growth, but may pose a significant threat to primary consumers. Zn is an essential plant nutrient but becomes toxic to vegetation at critical levels. Both metals have previously been shown to be readily taken up by poplar species [2, 3].

An understanding of the uptake, compartmentation and toxicological effects of metals to trees and other vegetation is imperative if greenspace establishment on contaminated sites is to be successful and to pose no significant risk to other receptors.

This paper presents simple models of metal uptake, by one variety of poplar, which were developed based on metal spiked hydroponic trials. The validity of these models is tested using results from soil based nursery and field trials. The data will be used to indicate whether the uptake of a metal into the biomass compartments of poplar can be predicted from the bioavailable or leachable fraction of metals in the soil.

2 Materials and methods

2.1 Hydroponics trial

The hydroponics data form part of a wider study in which the metal uptake of several varieties of poplar and willow were modelled in order to select the most suitable indicator species. This was found to be the poplar variety Fritzi Pauley (*Populus trichocarpa* Torr. & Gray). In the trials plant stock consisted of 20 cm unrooted cuttings. The poplar cuttings were planted in sealed one litre pots containing 900 cm^3 washed horticultural grade perlite. Perlite encourages good aeration and has a cation exchange capacity which is close to zero, resulting in an ion availability approaching 100% in solution. The pots were aerated with laboratory air throughout the experiment to ensure that aerobic conditions were maintained. Each pot also contained one of six solutions: control (¼ strength Hoagland's solution [4], or one of five Cd or Zn treatments in ¼ strength Hoagland's solution. Cd treatments were added as $CdSO_4.5H_2O$ to yield solutions of 0 (control), 1, 2, 3, 4, and 6 mg Cd l^{-1}. Zn treatments were added as $ZnSO_4.7H_2O$ to yield solutions of 0.02 (control), 5, 10, and 50 mg Zn l^{-1}. A fully replicated randomised factorial experimental design was set up in a glasshouse and harvested after 56 days. Trees were grown under 16 h of artificial light (PAR) and 8 h darkness per day. The temperature of the glasshouse was regulated to be between 20°C and 30°C. Every three days the volume of solution that had been transpired by the trees was calculated by measuring the mass lost from each pot and replenished with the appropriate solution.

2.2 Nursery trial

Poplar cuttings were grown in triplicate in pots containing 3.75 kg of a soil mixture for a period of two months. The soil had a silty-clay texture, was highly contaminated with Zn and also contained elevated concentrations of Cd (9580 mg kg^{-1} and 128 mg kg^{-1} respectively). The soil was amended, at a range of rates, with composted garden green waste or composted sewage sludge up to 20% (w/w). During the growing period, the water content of the soil in all pots was maintained between 0.35 and 0.5 m^3 m^{-3} with the pots being allowed to freely drain to prevent anaerobic conditions. After two months of growth, the stem, shoot, leaves and roots of the poplar were harvested for analysis. Soil mixtures from the nursery trial were leached for leachable metals using a 24 hour batch extraction with a demineralised water soil ratio of 10:1 [5]. After 24 hours the water fraction was filtered through a 0.45 μm filtration membrane and analysed for metals using ICP-OES.

2.3 Field study

Poplar cuttings were planted in a field study to assess whether the models developed using nursery data would hold true for data collected from plants growing in a contaminated soil *in situ*. Poplar cuttings were planted at the nodes of a 30 x 35 m grid that was divided into 42 blocks each measuring 25 m^2. The trial took place in an area of land contaminated as a result of former munitions manufacturing processes and contained heavy metal contaminants, in particular Zn (1040±1468 mg kg^{-1}). Analysis of the soil concluded that it was largely composed of industrial waste and foundry sand contaminated by metals. Soil samples were taken from the locations of each poplar cutting. After 12 weeks growth the cuttings were destructively harvested for analysis. Soil samples from the field study from each planting location were taken and bioavailable metals were determined using a 0.01M calcium chloride extraction followed by ICP-OES analysis.

2.4 Laboratory analyses

Plant samples were split into their leaf and shoot components. The foliar samples were dried at 70°C for 24 hours before grinding with a Cyclotec mill. Samples were dry-ashed at 500°C and digested with a concentrated hydrochloric and nitric acid mix (3:1 v/v). Resultant digests, leachate samples and calcium chloride extracts were analysed by ICP-OES (Spectro Analytical Instruments, West Midlands, UK) to determine concentrations of Cd and Zn.

2.5 Statistical data analysis

Data obtained throughout this study were subjected to generalised linear regression analysis to assess the significance of changes in observation with increasing Cd and Zn concentration, using the Genstat version 8.1 [6] statistics computer software package. Where necessary, data with skewed distributions were $\log_e$ transformed to comply with the assumptions of the regression model. Relationships between species and varieties were also studied using generalised linear regression analysis and t-probabilities of pairwise differences.

Linear and exponential models of the Cd and Zn uptake values were fitted using Genstat version 8.1 [6]. As the linear can be viewed as a limit of the exponential, the models were compared using the difference in residual sum of squares of alternative models relative to the smallest residual mean square to determine the more appropriate model. This comparison is referred to as an F-distribution with 1, n degrees of freedom where n is the residual degrees of freedom. When a model had been selected, the 95% confidence intervals were calculated.

3 Results

3.1 Metal compartmentation

The data suggest that Cd and Zn are compartmentalised in different ways within Fritzi Pauley (Figure 1). Whilst for both metals the concentrations in the stem tissue were lower than the other above ground plant compartments, a smaller proportion of Zn was present in the stem (10-22%) than Cd (18-28%). In addition, the shoot and leaf concentrations were broadly similar for Cd ranging between 30 and 43%, but for Zn a much higher proportion was present in the leaf tissue (49-61%) than in the shoot tissue (27-32%).

a) b)

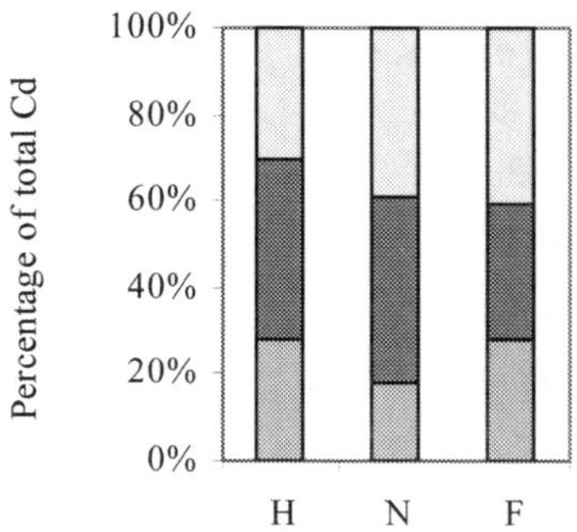

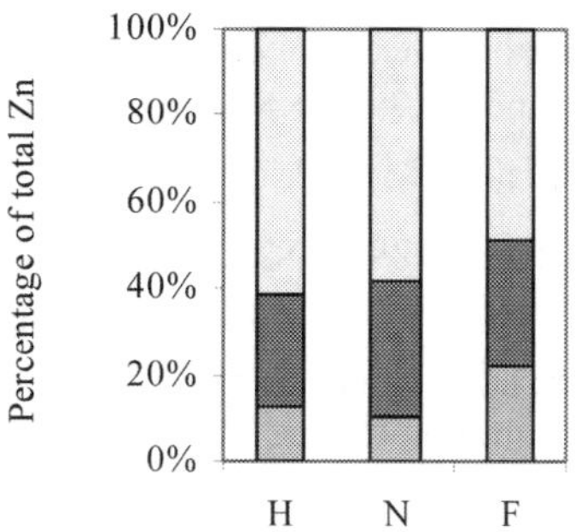

H= hydroponics; N=nursery; F=field; diagonal hatching=stem; vertical hatching=shoot; spots=leaf.

Figure 1: Metal compartmentation of Fritzi Pauley poplar cuttings grown in three different trials a) Cd; b) Zn.

The proportion of Cd in the stem was the same in the hydroponic and field trials (28%) and higher than in the nursery trial (18%). The percentage in the shoot tissue was similar in the hydroponic (40%) and nursery (43%) trials, being higher than in the field (32%); whilst the leaf proportions were similar in the nursery (39%) and field (40%) trials, being higher than the hydroponic trials (30%). The proportion of Zn in the shoot tissue was broadly similar across all experimental conditions (27-32%), as were those for stem (10-12%) and leaf tissue (58-61%) in the hydroponic and nursery trials. The higher percentage in stem tissue (22%) in the field trial is reflected in a lower leaf proportion (49%).

3.2 Nursery and field data

Linear regression analysis was used to determine whether the metal concentration in the leachates from the soil mixtures used in the nursery trial could be used to predict metal uptake into the plant tissues. The results show a significant positive relationship between the leachate concentration of Zn and the Zn uptake into leaf, shoot and stem of the poplar ($r=0.45$, $p<0.001$; $r=0.48$, $p<0.001$; $r=0.24$, $p<0.001$ respectively). Uptake of Cd into leaf and shoot showed no significant relation to the leachate concentration.

Linear regression analysis was used to determine whether the concentration of metals within plant tissues could be used to predict the bioavailable fraction of the soil contamination in the field. Analysis of Fritzi Pauley leaf metal concentrations showed statistically significant positive relationships with the bioavailable fraction of soil contamination for Zn ($r=0.33$, $p=0.002$) and Cd ($r=0.37$, $p=0.005$). Regression analysis performed with shoot metal concentration again showed statistically significant positive relationships for Zn ($r=0.27$, $p=0.004$) and Cd ($r=0.30$, $p=0.013$.

3.3 Metal models

Linear or exponential models were fitted to the data produced by the hydroponics experiment. These models can be expressed as:

$$\text{Cd Shoot} = (7.633 \times \text{Solution concentration}) + 3.85 \qquad r=0.87,\ p<0.001.$$

$$\text{Cd Leaf} = (5.960 \times \text{Solution concentration}) + 3.20 \qquad r=0.74,\ p<0.001.$$

$$\text{Zn Shoot} = 890.9\text{-}890.9\left(0.90^{\text{Solution concentration}}\right) \qquad r=0.84,\ p<0.001.$$

$r=0.86$, $p<0.001$.

where shoot and leaf concentrations are expressed as mg kg^{-1} and solution concentrations as mg l^{-1}.

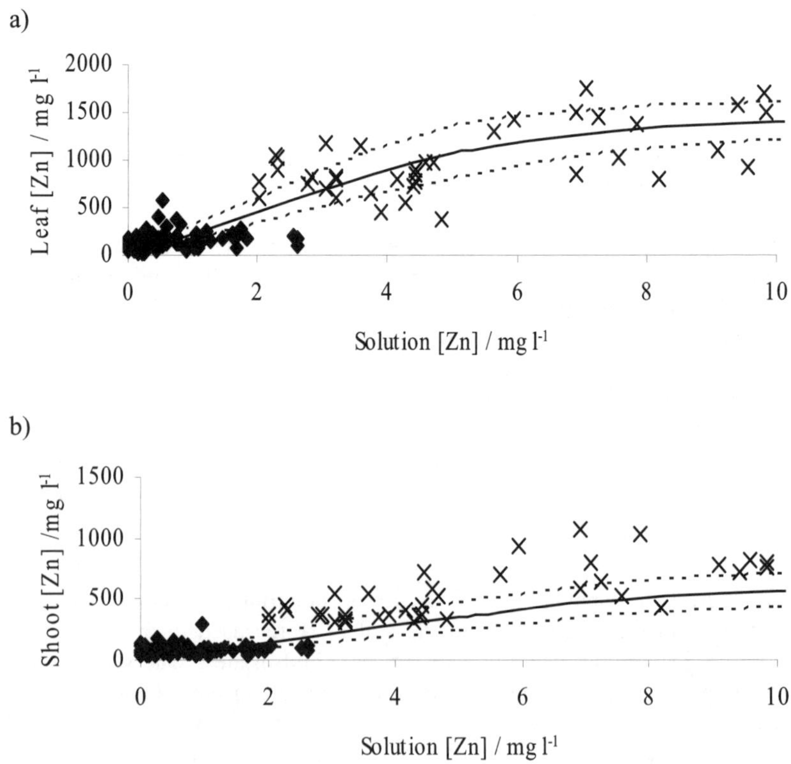

Figure 2: Models for Zn uptake into Fritzi Pauley a) leaf and b) shoot tissue. (solid line=model; dotted line=model 95% confidence intervals; x=nursery data; ♦=field data).

The significant data from the nursery and field trials were overlaid onto the models presented for Cd and Zn. Whilst the Cd values for the field trial fitted within the 95% confidence intervals for the models the values were so low that the models could not reliably be assessed. Similarly, the Zn data from the field trial represented the lowest end of the Zn range tested in the hydroponic trial (Figure 2), although they provided a better indication of the model's validity than for Cd. The Zn data from the nursery trial were generally in good agreement with the models (Figure 2). The majority of the points lay within the 95% confidence intervals for the leaf uptake model. The shoot values from the nursery trial were generally higher than those predicted by the model; the majority falling above the upper 95% confidence interval.

4 Discussion

The differences observed between the way in which Cd and Zn are compartmentalised within the above ground biomass of Fritzi Pauley may be due to the different requirements the tree has for these metals. Cd is not an essential nutrient; therefore the plant has no mechanism for controlling its compartmentation. Conversely, Zn is an essential micronutrient so the plant will direct this element to the tissue in which is it needed, hence the higher proportions in the leaves. Other workers have also found higher Zn concentrations in the leaves of poplar compared to other compartments [7–9]. The stem is used primarily as a transport mechanism for water and nutrients within the xylem and is therefore expected to have the lowest concentrations. The differences in the way these metals are accumulated in Fritzi Pauley has implications for the potential for food-chain transfer to primary plant consumers. The accumulation of both Cd and Zn in the leaf material, compared to the concentrations within the growth medium, suggest that fauna feeding on this material would be subjected to levels much higher than the available concentrations in soil. In addition, the cycling of contaminants through leaf fall may represent a significant risk to decomposers within the soil [10].

The differences in compartmentation between the trials may be explained by a number of factors. Firstly, the levels in solution in the hydroponic trial were considerably higher than both the nursery (measured as water extractable) and field trials (measured as $CaCl_2$ extractable). It is likely that the lower leaf proportions observed in the hydroponic trial may have resulted from a toxic effect of Cd, which may limit transpiration reducing further uptake [11]. The highest leaf concentrations measured in the hydroponic trials were approximately 30 mg kg^{-1} which are substantially higher than the normal range of 0.2-0.8 mg kg^{-1} [12], but well below the proposed upper critical concentration of Cd in leaf tissue of 200 mg kg^{-1} [1]. The similarities between the Zn compartmentation in the hydroponic and nursery trials may be due to the optimum growth conditions which the pots were grown under, in terms of light, temperature and water, than would have been experienced by the trees in the field. In addition, the higher proportions in leaf tissue may be because these trials both used Zn levels that resulted in tissue concentrations well above the

900 mg kg^{-1} critical value proposed by Dickinson [1]. The leaf concentrations in the field trial were well below this value, with a large majority falling within the 'normal' range for plant material of 8-400 mg kg^{-1} [12]. The Cd and Zn levels in the leaf tissue at the field plot were similar to those reported by Laureysens [2] for Fritzi Pauley grown in contaminated soil. Zn levels were also similar to those reported in the above ground biomass of *P. trichocarpa* and *P. nigra* grown on contaminated soil [13], but much higher than those reported for poplar grown on a tannery waste with a high soil organic matter content [14]. Cd levels were substantially lower than those reported for *P. trichocarpa* and *P. nigra* grown on contaminated soil with a significantly higher Cd concentration [13]. Different species will accumulate metals at different rates [8, 9, 13] and there is considerable variation between clones of the same species [2]. The poplars used in the field trial were *in situ* for a 12 week period which does not preclude the possibility of toxic levels being reached if they remained in the soil for longer time periods. It is the accumulation of metals in the leaves of vegetation over time that can result in harm to the trees and present a risk to consumers, and accumulation is likely to increase with the age of the trees [9].

The Cd and Zn levels used in the hydroponic study were above what would normally be expected in a soil solution, but were chosen to represent the range between what is 'normal' and the worst-case scenarios at contaminated sites. The mobile proportion of Cd and Zn in a contaminated soil is usually expected to be in the range of 0.5-3.2% and 4-18% respectively [13]. The p values for the models show a significant relationship between metal in solution and the metal concentrations in the tissue, and the r values are all high (> 0.74). The internal leaf concentrations in this study are in the same order of magnitude, but slightly higher than those reported in hydroponically grown red maple, white pine and Norway spruce [8]. They are also higher than those observed in a study using the poplar variety I-214, which had leaf concentrations of 131 mg kg^{-1} of Zn after only two additions of 1000 μM solution [15].

There are a number of problems with applying data obtained from hydroponic studies to site conditions. Soil properties such as the chemical form of the metal, competing metal ions and soil pH, CEC and organic matter will all strongly influence the availability of the contaminants to vegetation [1]. The ability of a tree to tolerate high metal levels may also vary between the nursery and field due to increased stresses in this environment. As mentioned above, trees grown in the nursery are done so under optimum conditions; the presence of these or additional stressors often found at former industrial sites, such as low organic matter or poor soil structure, may reduce metal tolerance in the field. In addition, hydroponically grown trees will not benefit from the presence of mycorrhizal fungi that may increase metal tolerance in the field [16].

Significant correlations were found between the bioavailable fraction of soil contamination and the concentration of metals within the leaves of the poplar cuttings. Calcium chloride is often used to determine the plant available fraction of soil [1, 2, 17]. The success of using this extractant is often dependant on the properties of the soil being analysed [1]. The weak relationships ($r<0.37$) between the amount of metal extracted by calcium chloride (0.01 M) and the measured plant uptake in the field study suggest that this method is not always a reliable way of assessing the potential for plant uptake.

The data for Cd from the nursery trial could not be overlaid onto the uptake models developed from the hydroponic work. Although, the field study data fitted into the 95% confidence intervals for the models, the Cd values were too low to realistically assess the reliability of the models. In contrast, the models for Zn uptake into the leaves and shoots can be validated using both the field and nursery data. The field leaf data are evenly spread around the base of model and the majority of the nursery results fitted within the 95% confidence interval of the model. Zn uptake in the shoots is similar for field results but the nursery results are only partly within 95% confidence interval, tending to show increased uptake compared to the hydroponic study. In summary, $CaCl_2$ extractable soil metals give significant but weak correlations against metal uptake into the leaves of poplar. However, this study demonstrates that it may be possible to use leaf concentrations of Fritz Pauley poplar grown as a bioindicator to assess the plant available concentration of zinc in soil. Such data can then be applied to models derived from hydroponic studies of metal uptake for other tree species to predict their likely survival, growth, and metal uptake characteristics and associated risk of food chain transfer.

5 Conclusion

The application of poplars as indicators of the plant available fraction of soil may provide a valuable tool to be used in conjunction with standard soil analysis techniques. The application of models to predict the available Zn concentration in soil using leaf concentrations looks promising. In contrast, the models for Cd need to be compared to Fritzi Pauley grown on soils with higher available levels of Cd before their validity can be truly assessed.

References

[1] Dickinson, N.M., Mackay, J.M., Goodman, A. & Putwain, P., Planting trees on contaminated sites: Issues and guidelines. *Land Contamination and Reclamation,* **8(2)**, pp. 87-101, 2000.

[2] Laureysens, I., Blust, R., de Temmerman, L., Lemmens, C. & Ceulemans, R., Clonal variation in heavy metal accumulation and biomass production in a poplar coppice culture: I. Seasonal variation in leaf, wood and bark concentrations. *Environmental Pollution,* **131**, pp. 485-494, 2004.

[3] Madejón, P., Marañón, T., Murillo, J.M. & Robinson, B., White poplar (*Populus alba*) as a biomonitor of trace elements in contaminated riparian forests. *Environmental Pollution,* **132**, pp. 145-155, 2004.

[4] Hoagland, D.R., & Arnon, D.I., The water-culture method for growing plants without soil. *Miscellaneous Publications No. 3514.* California Agricultural Experiment Station, California, 1941.

[5] BSI, Characterisation of waste - Leaching - Compliance test for leaching of granular waste materials and sludges. *BS En 12457: Part 2*, British Standards Institution, London, 2002.

[6] GenStat, *The Guide to GenStat Release 8.1 Part 2: Statistics*, ed. R.W. Payne, Lawes Agricultural Trust, Rothamsted Experimental Station, VSN International: Oxford, 2005.

[7] Pulford, I.D. & Watson, C., Phytoremediation of heavy metal-contaminated land by trees – a review. *Environment International,* **29**, pp. 529-540, 2003.

[8] Mitchell, C.D. & Fretz, T.A., Cadmium and zinc toxicity in white pine, red maple and Norway spruce. *Journal of the American Society of Horticultural Science*, **102(1)**, pp. 81-84, 1977.

[9] Hermle, S., Günthhardt-Goerg, M.S. & Schulin, R., Effects of metal-contaminated soil on the performance of young trees going in model ecosystems under field conditions. *Environmental Pollution*, in press.

[10] Hobbelen, P.H.F., Koolhaas, J.E. & van Gestel, C.A.M., Effects of heavy metals on the litter consumption by the earthworm *Lumbricus rubellus* in field soils. *Pedobiologia*, **50**, pp. 51-60, 2006.

[11] Lamoreaux, R.J. & Chaney, W.R., Growth and water movement in silver maple seedlings affected by cadmium. *Journal of Environmental Quality*, **6(2)**, pp. 201-205, 1977.

[12] Ross, S. M., Sources and forms of potentially toxic metals in soil-plant systems. *Toxic Metals in Soil-Plant Systems*, ed. S.M. Ross, John Wiley & sons: Chichester, pp. 2-27, 1994.

[13] Fischerová, Z., Tlustoš, P., Száková, J. & Šichorová, K., A comparison of phytoremediation capability of selected plant species for given trace elements. *Environmental Pollution*, in press.

[14] Giachetti, G. & Sebastiani, L., Metal accumulation in poplar grown with industrial wastes. *Chemosphere*, in press.

[15] Di Baccio, D., Tognetti, R., Sebastiani, L. & Vitagliano, C., Responses of *Populus deltoides* x *Populus nigra* (*Populus x euramericana*) clone I-214 to high Zn concentrations. New Phytologist, **159**, pp. 443-452, 2003.

[16] Kim, *C.-G.*, Power, S.A. & Bell, J.N.B., Effects of host plant exposure to Cd on mycorrhizal infection and soluble carbohydrate levels of *Pinus sylvestrus* seedlings. *Environmental Pollution*, **131**, pp. 287-294, 2004.

[17] Young, S.D., Zhang, H., Tye, A.M., Maxted, A., Thums, C. & Thornton, I., Characterising the availability of metals in contaminated soil. I. The solid phase: sequential extractions and isotopic dilution. Soil Use and Management, **21**, pp. 450-458, 2006.

Section 4
Cleanup methodologies

Black alder as a promising deciduous species for the reclaiming of oil shale mining areas

K. Lõhmus[1], J. Truu[2], M. Truu[2], E. Kaar[3], I. Ostonen[1], S. Alama[1], T. Kuznetsova[3], K. Rosenvald[1], A. Vares[3], V. Uri[3] & Ü. Mander[1]
[1]*Institute of Geography, University of Tartu, Estonia*
[2]*Institute of Molecular and Cell Biology, University of Tartu, Estonia*
[3]*Institute of Forestry and Rural Engineering, Estonian University of Life Sciences, Tartu, Estonia*

Abstract

Black alder (*Alnus glutinosa*) plantations of different age were investigated on reclaimed oil shale mining areas in Northern Estonia with the aim of analysing the impact of microbial communities in the rhizosphere and bulk soil, and the effect of fine root adaptations on stand development and productivity. Biolog Ecoplates were used to determine the community-level physiological profiles (CLPP) in the rhizosphere and bulk soil samples. CLPP were summarized as AWCD (average well color development, OD $48h^{-1}$). The rhizosphere/bulk soil ratio of AWCD was an order higher in the 4-yr-old stand than in the first year after planting or in middle-aged stands. The substrate-induced respiration (SIR) and basal respiration (BAS) of bulk soil samples were measured, and the metabolic quotient $q(CO_2)$=BAS/SIR was calculated. Short root morphological studies were carried out using WinRHIZO™ Pro 2003b. SIR increased from 0.23 to 2.73 $mgCg^{-1}$, while $q(CO_2)$ and mean specific root length (SRl, m g^{-1}) decreased with increasing stand age (from 1 to 26 years) from 0.51 to 0.25, and from 172 to 90 m g^{-1} respectively. Soil pH decreased 1 unit during the first 26 years. According to efficient adaptive strategies, the survival and productivity of black alder stands on oil shale mining areas are high, and hence black alder is a perspective tree species for the afforestation of these areas.
Keywords: Alnus glutinosa, Biolog Ecoplates, fine root adaptations, oil shale mining area, rhizosphere processes, substrate-induced and basal respiration.

WIT Transactions on Ecology and the Environment, Vol 94, © 2006 WIT Press
www.witpress.com, ISSN 1743-3541 (on-line)
doi:10.2495/BF060091

1 Introduction

1.1 Reclamation of disturbed landscapes in Northern Estonian oil shale mining area

Every year opencast mining in the Northeast Estonian oil-shale field – the largest commercially exploited oil-shale deposit in the world (>6600 million tons) – creates substantial areas of wasteland. Presently, 2 underground mines and 2 open pit mines are in operation. In this area the overburden reaches ranges of 0-70 m due to the gentle southward dipping of the Ordovician strata. Mining in pits began to spread intensively in 1959. Pits are used in digging up to depth of 40 m; if oil shale is deeper than 50 m underground, mining is used. Nowadays around 50% of oil shale is mined in pits. The relief of the alkaline (pH~8) wasteland is rugged, the soil heterogeneous and extremely stony. Stone content varies from 15 to 100%; the N and organic content of oil shale mining spoil is low. Hence, afforestation is an optimal tool for the reclamation of these disturbed landscapes, as well as being a sustainable management option to create renewable energy sources.

The extensive afforestation of exhausted opencast oil-shale mines in Northeast Estonia was begun in 1960, and on the 1st of January 2005, exhausted oil-shale mines covered 12,900 ha of land, of which 10,200 ha had been forested. Until recent years a disproportionately large proportion of conifers (>90%), mainly Scots pine (*Pinus sylvestris*), has been planted (86% of the area). A total of 52 indigenous and introduced woody species were planted, where different introduced species of *Larix* (*Larix europaea*, *L. sibirica*, *L. kurilensis*) showed the best growth among coniferous trees, and the native deciduous species silver birch (*Betula pendula*) and black alder (*Alnus glutinosa*) were the most successful deciduous trees. They exceed pine stands of the same age (25-30 years) by 4 to 7 metres in height. The recommended planting density for black alder is 2000 - 2500, and hence thinning at a young age is unnecessary. In very stony areas (stones compromise 50-70%), black alder. is more suitable to promote soil formation processes and surpassed the growth of silver birch.

1.2 Advantages of alders among woody species

Alders have a number of advantages over other cultivated species: the increased N and P availability in the soil under alders, faster growth at a young age and higher resistance to pests, diseases and fires, especially in comparison with conifer monocultures. Owing to the ability to fix N_2 by the symbiosis of actinomycetes *Frankia* in alder root nodules, the soil under alders is enriched in nitrogen. Due to the low nitrogen retranslocation from senescing leaves (from 2.5 to 14%), alder leaf litter is extremely rich in nitrogen and mineralises easily [1, 2]. Alders increase phosphorus availability in the soil through the activity of their roots and associated microbial communities [3, 4]. Thus alders improve both soil N and P status. The majority of alder roots with primary structure are ectomycorrhizas, and hence mycorrhizal symbiosis plays an important role in the nutrient cycling of alder stands. The nutrient demand of alders is higher and N use efficiency is lower than that of other tree species. In order to satisfy the high demand for nutrients, beneficial rhizosphere conditions should be supported by alders [5].

1.3 Rhizosphere processes and fine-root adaptations in black alder

The potential of black alder for the recultivation of exhausted opencast oil shale mines in relation to fine-root adaptations and rhizosphere processes is still poorly understood. Populations of microbes in the rhizosphere differ quantitatively and qualitatively from those in the bulk soil; their numbers are generally higher, and different populations are commonly represented [6]. It has been shown that, in black alder, the size of the total microbial population and the numbers of ammonifying and proteolytic microorganisms are higher in the soil-root interface than in bulk soil, litter, and root-free soil under trees [7]. Although rhizodeposition strongly affects the structure and activity of soil microbial communities and plant nutrition, few studies focus on the root surface bacteria of forest trees. The interaction of roots and soil microbial communities should be especially important in harsh site conditions, including reclaimed opencast oil-shale-mining areas [5].

Biolog Ecoplates, where all substrates are known as root exudates, were used to determine the community-level physiological profiles (CLPP) of culturable bacteria in soil-root interface and bulk soil samples [8]. The method characterizes the part of culturable bacteria and not fungi, but in studied alder sites the fungal/bacterial ratio [9] should be low, due to the soil pH_{KCl} range (from 7.0 to 8.0). An advantage of Biolog microplates is that they allow one to assess the functional diversity of the culturable microbial community. It is assumed that higher activity and diversity values for culturable microbial communities correspond to respectively higher values of total rhizosphere communities and can be used as indicators of their activity and diversity. In order to estimate the soil microbial biomass and metabolic activity of humus-degrading microorganisms present in the bulk soil, respiration techniques SIR and BAS were measured [10–12].

In considering microbial communities in the rhizosphere, the morphological and functional variability of fine roots is important. The impact of fine-root morphological parameters on rhizosphere processes in trees has not yet been thoroughly investigated. We measured short-root size and functional parameters: specific root area (SRA) and specific root length (SRL) [5, 13, 14] in alder stands. The results of different approaches: measurements of the size and activity of microbial communities by respiration in soil, the assessment of culturable bacterial communities with Biolog Ecoplates, the foliar assimilation efficiency of the above-ground part of trees, and short root morphology, especially the functional parameter SRA, were in close accordance [5].

The impact of alder species on fine-root morphology was significant, and short root tips were larger for black alder than or grey alder. For the investigated microbiological characteristics, no alder-species-related differences were revealed. The AWCD values of culturable bacterial communities in the soil-root interface compared to those in the bulk soil correlated positively with specific short-root area and negatively with foliar assimilation efficiency [5]. The first years of stand development are most critical for tree survival, and it is for that reason that one-, four- and 27-yr-old stands were included in the study. The higher growth rate of black alder corresponds to the higher activity of microbial communities in their rhizosphere, and the higher specific root area of short roots than in the case of conifers [5, 14, 15].

1.4 Aims and objectives of study

(1) To analyse the impact of microbial communities in the rhizosphere and in the bulk soil, and of fine-root adaptations on the development and productivity of black alder stands on reclaimed oil-shale mining areas.
(2) To make recommendations concerning the suitability of black alder for the recultivation of exhausted oil-shale opencast mines.

2 Materials and methods

2.1 Stand and soil characteristics

A one-yr-old black alder plantation (Narva II, 3 replication plots) established in 2005 and a 4-yr-old plantation (Narva I) established in 2002 on oil shale mining spoil were investigated in October 2005 and 2004 respectively. A middle-aged Sirgala stand was established in 1978 [15] the present soil type is *Spolic Anthrosol* [14], and the stand was involved in the study in October 2002 and 2004. The planting arrangement was in all cases 2x2 m, bare-root seedlings were one year old in Sirgala and Narva I, and in Narva II were two years old. Stand and soil characteristics are presented in Table 1, initial N and organic content is low, and in young stands a significant proportion of the organic matter is formed from oil shale mining residues [16].

Table 1: Stand and soil characteristics in 2005. DBH (cm) – diameter at breast height (*indicates diameter at root collar); H- tree height (m), N% and LOI% indicate the percentage of nitrogen and loss on ignition respectively.

Stand	Age	Trees per ha	DBH	H	N%	LOI%
Sirgala	27	1650	14.8	20.3	0.55	15.0
Narva I	5	2100	*2.8	0.98	0.04	1.3
Narva II	1	2300	*0.8	0.33	0.03	4.7

2.2 Soil sampling and processing

Ten samples from the 0-10 cm soil layer (20x20 cm^2) were taken randomly in the Sirgala black alder stand in October in 2002 and 2004. One third of the root system was taken from the 4-yr-old Narva II stand, and the whole root system with the soil of 10 randomly selected trees was taken from the one-year-old Narva I stand (from all 3 replicate plots) in October 2004 and 2005, respectively. A compound subsample per plot was processed according to the methodology proposed by Gobran and Clegg [17]. Gobran and Clegg [17] introduced a conceptual model for nutrient availability in the mineral soil-root system, in which the fine roots and associated organisms maintain a higher nutrient availability in the soil-root interface than in the bulk soil.

All roots were carefully removed by hand from the field-moist mineral soil, which was then passed through a 2-mm mesh sieve to yield the bulk soil fraction. The dead roots, nodules, and coarse roots (≥ 2 mm in diameter) were separated; the remaining fine roots ($d < 2$ mm) and soil were gently shaken for 1 min in a plastic container to separate the soil aggregates from the roots. The remaining fine roots with adhering soil yielded the rhizosphere fraction.

2.3 Chemical analysis

Nitrogen in the rhizosphere and bulk soil samples was determined using the Kjeldahl method with a Tecator ASN 3313. Loss on ignition (LOI) was determined at 360°C; the pH_{KCl} of samples was measured. Analyses were performed at the Biochemistry Laboratory of the Estonian University of Life Sciences.

2.4 Microbiological methods

Biolog Ecoplates (Biolog Inc.) where all carbon sources are known as root exudates were used to determine the community-level physiological profiles of culturable bacterial samples in the soil-root interface and bulk soil; 1g of fresh material was used in all cases, and the data were recalculated on the basis of the dry matter. Biolog profiles were summarized as AWCD (average well color development). AWCD - average well color development; this is the sum of all 31 substrates' utilization values by culturable bacteria, divided by 31. A 150-µl aliquot of a 10^{-4} dilution of the bulk soil or soil-root interface sample was added to each of the 96 wells (31 carbon sources and control in 3 replications) in the microplate. Plates were incubated at 25°C and colour development was measured every 24 h for 120 h as absorbance at 590 nm, with optical density plate reader Multiscan 340 C. Based on the examination of the kinetic curves of the AWCD, 48h measurements were chosen for further data analysis.

Active microbial biomass was determined using substrate-induced respiration (SIR) [10–12]. Microbial respiration activity (BAS) was measured by trapping the evolved carbon dioxide in sodium hydroxide. The carbon availability index, also called metabolic quotient $q(CO_2)$, which relates the respiration rate without added substrate (BAS) to respiration after the addition of sufficient readily available substrate (SIR), was also calculated.

2.5 Morphological parameters of fine roots

The fine-root parameters of black alder were measured in 10 samples per year and plot, and 2-3 subsamples were processed per sample. Only short roots with a living cortex were considered: the diameter, length, volume and mass of root tips, as well as specific root area and specific length, were estimated. The methods are the same as in [13], except for the diameter, root length and root projection area measurements, which in the present work were carried out using WinRhizo 2003B (Regent Instruments) [14].

2.6 Statistical methods

Kolmogorov-Smirnov, Lilliefors and Shapiro-Wilk's tests were used to check the normality of variables. When necessary, log- and root-transformations were used to normalize the data. Differences between stand means of short-root characteristics were checked at 95% confidence intervals. The level of significance of α= 0.05 was accepted in all cases. The STATISTICA 7.0 software was used.

3 Results and discussion

3.1 Strategies to optimise mineral nutrition

To sustain and improve mineral nutrition in the harsh conditions of the levelled hills of oil shale mining spoil, trees must invest assimilates according to an extensive or intensive strategy. Concerning the optimality theory, the cost/benefit ratio of operating roots is a key factor for forest productivity [18].

There are two main strategies to optimise the mineral nutrition of plants:
A) Extensive, by increasing the mass, surface area and length of fine roots, leading to the increase of the rhizosphere.
B) Intensive, by increasing or maintaining the efficiency of fine roots and rhizosphere processes through morphological adaptations of fine roots [13, 19], and/or the activity of root-associated microorganisms for plant mineral nutrition could be increased or maintained.

Hence an extensive strategy leads to an increase in fine root system, but an intensive strategy is based on more efficient use of the mass unit of ectomycorrhizas.

3.2 Dynamics of fine root adaptations and pH in black alder stands

The mean specific root area of ectomycorrhizas was 171±12, 155±8, and 82±4 $m^2\ kg^{-1}$ in one-, four- and 27-yr-old black alder stands respectively (Table 2), whereas in the 27-yr old stand the value was significantly higher.

The mean diameter and weight of a short root tip were significantly smaller in a one-yr-old plantation than in older stands, most probably as a result of planting shock. Mean specific root length decreased significantly in correlation with increasing stand age (Fig.1A).

Phosphorus nutrition was improved through the increasing of rhizosphere pH by up to 1.4 units compared to the bulk soil; soil pH decreased 1 unit during the first 26 years Fig. 1B).

Table 2: Mean short-root morphological parameters (± standard errors) in black alder stands. Different letters indicate significant differences between means by 95% confidence intervals, P<0.05.

Stand	Diameter (mm)	Length, (mm)	Weight, (mg)	SRA, (m^2 kg^{-1})
Sirgala 27 yr	0.291^a±0.005	3.02^a±0.18	0.035^b±0.003	82^a±4
Narva I 4 yr	0.392^c±0.005	4.10^b±0.19	0.034^b±0.002	155^b±8
Narva II 1 yr	0.321^b±0.008	3.51^a±0.38	0.021^a ± .002	171^b±12

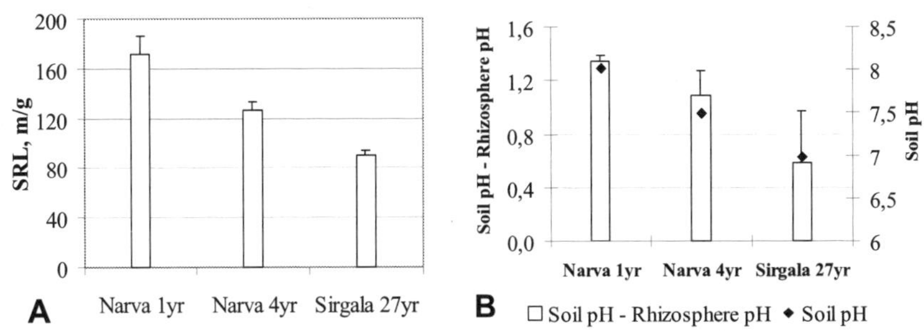

Figure 1: A: Decrease in mean SRL of short roots with increasing stand age; B: Acidifying effect of rhizosphere and soil pH(KCl) dynamics in black alder stands on reclaimed oil shale mining area. Bars indicate standard errors.

3.3 Activity dynamics of microbial communities in rhizosphere and soil

Oil shale mining spoil is extremely low in nitrogen and organic matter. Additionally, the microbial abundance is initially very low, as can be seen in Fig. 2A. The soil improvement during the first 25 years is remarkable (Table 1, Fig.1B and 2A).

The metabolic quotient $q(CO_2)$ was similar in two young stands, decreasing approximately twice during stand development (Fig. 2B) This reflects successional changes in microbial community structure, where in the soil of the young stands the microbial community consists mainly of r- strategist species that rapidly utilize root derived organic substrates and are not as efficient in the formation of biomass.

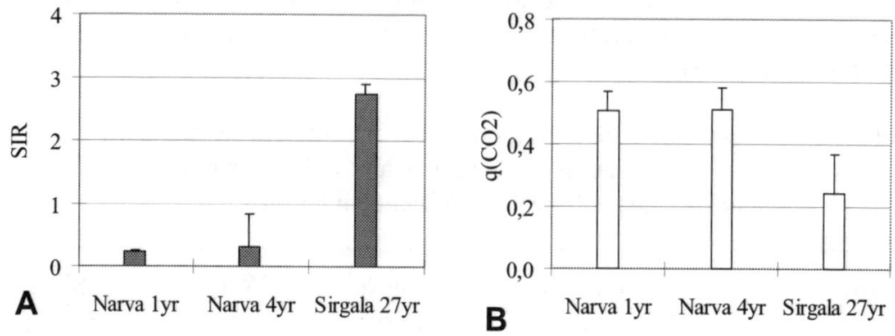

Figure 2: A: Dynamics of SIR (mg C g^{-1}); B: Decrease in metabolic quotient $q(CO_2)$ with increasing age of black alder stands on reclaimed oil shale mining area. Bars indicate standard errors.

In older stands where plant litter has been accumulating, the microbial community also contains K-strategic species that are characterized by a slow growth rate and stable biomass development.

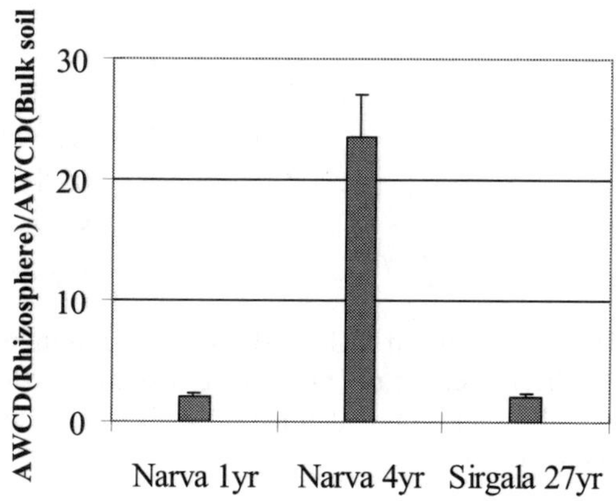

Figure 3: Rhizosphere /bulk soil AWCD ratio in black alder stands on reclaimed oil shale mining area. Bars indicate standard errors.

The greatest difference between one-, 4- and 27-yr-old stands was revealed in the rhizosphere/bulk soil AWCD ratio, which reflects the difference between activities of rhizosphere and bulk soil microbial communities (Fig. 3). Hence, in the second year after planting, when survived plants have overcome planting shock, the intensive strategy prevails, and plants support rhizosphere microbial communities massively. The rhizosphere/bulk soil activity ratio is an order higher than in the first year after planting or in a middle-aged stand. In our earlier investigation it was revealed that the better the soil conditions, the smaller the difference between the activities and diversities of microbial communities in the rhizosphere and soil. Hence, by improved soil the support of rhizosphere communities is less crucial.

Why are the rhizosphere communities poorly supported in the first year after planting, when the need is most urgent? It seems that during the first year the intensive strategy of the development of the fine root system prevails in order to exploit the oil shale mining spoil by fine roots as much as possible, and there is a deficit in assimilates allocated below ground. That priority is reflected by the highest short root SRL being found in the one-yr-old stand (Fig. 1A), although the SRA of ectomycorrhizas is similar in one-yr-old and 4-yr-old stands. That is in close accordance with the low support of rhizosphere microbial communities, because in our earlier investigation the Rhizosphere /bulk soil AWCD ratio, ratio for Shannon diversity indices, and SRA were positively correlated ($r=0.65$, $P<0.05$) in both cases [5].

In Sirgala, the analysis of topsoil nutrients showed a remarkable increase in total N and available P under the canopy of the black alder plantation in comparison to the neighboring Scots pine stand of the same age. A more than twice higher total topsoil N and available P content were found in the alder plantation compared to the pine stand [15]. The above-ground productivity in the 21-year-old black alder plantation on the reclaimed oil-shale mining area in Sirgala was comparable with the value for stands of the same age growing on fertile mineral soils. The biomass accumulation ratio, foliar assimilation efficiency and N use efficiency were highest in the Sirgala plantation. In Sirgala, the planting density ensured nearly optimal photosynthetic conditions, and the planting of 2,000 to 2,500 black alders per hectare on exhausted opencast oil-shale mines was recommended.

4 Conclusions

Black alder grows rapidly on recultivated oil shale areas; its survival after establishment is very good. A favorable environment for microbes in rhizosphere and bulk soil is created on oil shale mining detritus under black alders. The morphological adaptations of ectomycorrhizas support the effective functioning of the developing ecosystem. Therefore black alder is a very promising tree species for the recultivation of exhausted opencast oil shale mines.

Acknowledgements

This study was supported by Target Funding Project No. 0182534s03 of the Ministry of Education and Science of Estonia and Grant No. 6011 from the Estonian Science Foundation. We would like to thank Vahur Kurvits for the help in field work and Mr. Alexander Harding for the linguistic revision of the manuscript.

References

[1] Dawson, J.O. & Funk, D.T., Seasonal change in foliar nitrogen concentration of *Alnus glutinosa*. *Forest Science*, 27, pp. 239-243, 1981.

[2] Lõhmus, K., Kuusemets, V., Ivask, M., Teiter, S., Augustin, J., & Mander, Ü., Budgets of nitrogen fluxes in riparian grey alder forests. *Archiv für Hydrobiologie. Suppl.* 141, pp. 321-332, 2002.

[3] Giardina, C.P., Huffman, S., Binkley, D. & Caldwell, B.A., Alders increase soil phosphorus availability in a Douglas-fir plantation. *Canadian Journal of Forest Research,* 25, pp. 1652–1657, 1995.

[4] Uri, V., Tullus, H. & Lõhmus, K., Biomass production and nutrient accumulation in short-rotation grey alder (*Alnus incana* (L.) Moench) plantation on abandoned agricultural land. *Forest Ecology and Management*, 161, pp. 169-179, 2002.

[5] Lõhmus, K., Truu, M., Truu, J., Ostonen, I., Kaar, E., Vares, A., Uri, V., Alama, S. & Kanal, A. Functional diversity of culturable bacterial communities in the rhizosphere in relation to fine-root and soil parameters in alder stands on forest, abandoned agricultural and oil-shale mining areas. *Plant and Soil*, (In press) 2006.

[6] Butler, J.L., Williams, M.A., Bottomley, P.J., & Myrold, D.D., Microbial community dynamics associated with rhizosphere carbon flow. *Applied Environmental Microbiology*, 69, pp. 6793-6800, 2003.

[7] Gonzales, J.M.P., Manero, F.J.G., Probanza, A., Acero, N. & Decastro, F.B., Effect of alder (*Alnus glutinosa* L. Gaertn.) roots on distribution of proteolytic, ammonifying, and nitrifying bacteria in soil. *Geomicrobiology Journal*, 13, pp. 129–138, 1995.

[8] Preston-Mafham, J., Boddy, L. & Randerson, P.F., Analysis of microbial community functional diversity using sole-carbon-source utilisation profiles – a critique. *FEMS Microbiology Ecology*, 42, pp.1-14, . 2002.

[9] Bååth, E. & Anderson, T.H., Comparison of soil fungal/bacterial ratios in a pH gradient using physiological and PLFA based methods. *Soil Biology and Biochemistry*, 35, pp. 955-963, 2003.

[10] Anderson, J.P.E. & Domsch KH., A physiological method to the quantitative measurement of microbial biomass in soils. *Soil Biology and. Biochemistry*, 10, pp. 215-221, 1978.

[11] Cheng, W., Zhang, Q., Coleman, D.C., Carroll, C.R. & Hoffmann, C.A., Is available carbon limiting microbial respiration in the rhizosphere? *Soil Biology and. Biochemistry*, 28, pp., 1283-1288, 1996.

[12] Schinner, F., Öhlinger, R., Kandeler, E. & Margesin, R., *Methods in Soil Biology*, Springer-Verlag: Berlin, 426 pp. 1996.

[13] Ostonen, I., Lõhmus, K. & Lasn, R., The role of soil conditions in fine root ecomorphology in Norway spruce (*Picea abies* (L.) Karst.). *Plant and Soil*, 208, pp. 283-292, 1999.

[14] Ostonen, I., Lõhmus, K., Alama, S., Truu, J., Kaar, E., Vares, A., Uri, V.& Kurvits, V., Morphological adaptations of short roots in Scots pine, silver birch and black alder stands in recultivated opencast oil shale mining and semi-coke areas. *Oil Shale*, 23(2), (In press), 2006.

[15] Vares, A., Lõhmus, K., Truu, M., Truu, J., Tullus, H. & Kanal, A., Productivity of black alder (*Alnus glutinosa* (L.) Gaertn.) plantations

established on reclaimed oil shale mining spoil and mineral soils in relation to rhizosphere conditions. *Oil Shale*, 21, 47-62, 2004.

[16] Reintam, L., Kaar, E., & Rooma, I., Development of soil organic matter under pine on quarry detritus of open-cast oil shale mining, *Forest Ecology & Management*, 171, pp. 191-198, 2002.

[17] Gobran, G. & Clegg, S., A conceptual model for nutrient availability in the mineral soil–root system. *Canadian Journal of Soil Science*, 76, pp. 125–131, 1996.

[18] Eissenstat, D., M. & Yanai, R.D., The ecology of root life span. *Advances in Ecological Research*, 27, pp. 1-62, 1997.

[19] Leuschner, C., Hertel, D., Schmid, I., Koch, O., Muchs, A. & Hölscher, D., Stand fine root biomass and fine root morphology in old-growth beech forests as a function of precipitation and soil fertility. *Plant & Soil*, 258, pp. 43-56, 2004.

The effect of composts on the leaching of metals from contaminated soils

R. van Herwijnen[1, 3, 4], T. Laverye[2], S. K. Ouki[1], A. Al-Tabbaa[3], M. E. Hodson[2] & T. R. Hutchings[4]
[1]*School of Engineering, University of Surrey, Guildford, UK*
[2]*Department of Soil Science, School of Human and Environmental Science, University of Reading, UK*
[3]*Department of Engineering, University of Cambridge, UK*
[4]*Land Regeneration and Urban Greening Group, Forest Research, Farnham, UK*

Abstract

Compost is often proposed as a suitable material for the remediation of contaminated brownfield sites intended for soft end-use. In addition to vitalising the soil, it is also supposed to immobilise metals and so break contaminant-receptor pathways and reduce the ecotoxicity of the contaminants. However, some research has demonstrated contradictory effects between composts on metal immobilisation. In our study, five different composts were tested to examine both their metal retention and toxicity reduction capabilities on three different metal contaminated soils. Leaching tests, a plant growth test with Greek cress (*Lepidium sativum*), an earthworm (*Eisenia fetida*) survival and condition test and a bacterial toxicity test using *Vibrio fischeri* were carried out. The batch leaching tests results showed that spent mushroom compost caused an increase in metal concentration in the leachates while LimeX70 compost caused a decrease. The variation in behaviour between different composts for each soil was somewhat random, so a generic conclusion could not be drawn. Toxicity tests showed significant reduction of metal bioavailability and toxicity. Our results also suggest that more rigorous tests should be undertaken to understand the mechanisms involved in metal complexation using different compost types, in order to optimise the use of compost for remediation.
Keywords: compost, heavy metals, toxicity, leaching, remediation.

WIT Transactions on Ecology and the Environment, Vol 94, © 2006 WIT Press
www.witpress.com, ISSN 1743-3541 (on-line)
doi:10.2495/BF060101

1 Introduction

Most large cities in the world have to deal with brownfield sites within their boundaries. Redevelopment of these sites would reduce the need for greenfield development and so protect greenfield land. The UK Environment Agency has estimated that around 20,000 contaminated sites in England and Wales may need treatment [1]. The costs of treatment of these sites can be made more bearable if their remediation is combined with redevelopment. Nevertheless developers generally prefer to choose greenfield land rather than brownfield land because of the high costs of the remediation of contaminated land [1]. This, and the recent increase in landfill costs for contaminated soils, has created a market for low cost alternative remediation techniques. One possible remediation solution for a brownfield site is its transformation into green spaces in the urban environment. Many urban brownfield sites have metal concentrations that significantly exceed the levels that are generally stated as environmentally acceptable and therefore form a potential health risk for humans, animals and plants. The threat of these metals to the environment can be reduced by fixation in the soil itself, thereby lowering the bioavailability and risk of further mobility. A considerable amount of research has been performed on the addition of compost to immobilise metals [2]. Earlier research (van Herwijnen et al. unpublished results) revealed that not all composts immobilise metals in soils. Compost derived from sewage sludge increased leaching of Cd and Zn from a polluted soil while compost derived from garden greenwaste immobilised these metals. In order to further investigate this observation we have screened a larger set of composts on their metal immobilising capacity as well as their capability to reduce the toxicity of the contaminated soil. The tested composts in this study were green waste compost, LimeX70 (compost produced from by-products gained through refining sugar), spent mushroom compost, sewage sludge compost and coir compost (composted coconut husks).

In order to get a more general picture of their effect, the composts were tested in combination with three different kinds of metal polluted soil varying from soil with slightly elevated levels of Cu and Zn to soil highly contaminated with Zn or As. The soil-compost mixtures were also tested for their toxicity because mobility of the metals is only weakly related to toxicity due to the different species metals might appear in. This paper presents an overview on whether these compost mobilise or immobilise metals and whether they can reduce toxicity of the polluted soils.

2 Materials and methods

2.1 Materials

Three contaminated soils, Avonmouth (AM), Devon Great Consols (DGC) and Thamesmead (TM), and five composts, green waste compost (GWC), LimeX70 (LX), spent mushroom compost (SM), sewage sludge compost (SC) and coir compost (CC), were used. An uncontaminated soil, Kettering Loam, was also used in some experiments as a control measure. All soils and composts were air dried and sieved to ≤4mm before any experiments were conducted. The soils and composts were characterised for organic matter content, textural class, pH and total metal content (As, Cd, Cr, Cu, Pb and Zn).

2.2 Batch leaching test experiment

The batch test experiment was conducted following protocols set out by the European Standards Committee [3]. Soil and compost were combined in a mass ratio of 5:1 and placed in an acid washed polyethylene container. Deionised water was added to this mixture in a mass ratio 10:1. For each mixture, 500 cm^3 of deionised water was mixed with 50 g of the soil and compost. Four replicates of each mixture were used. The soil/compost suspensions were placed on a rotary shaker set at 10 rpm for 24 hours. After 24 hours the suspensions were filtered first through a 1.2 μm filter membrane (Whatman, cellulose nitric membrane) to remove most of the suspended sediment material, and then through a finer 0.45 μm filter membrane (Whatman, cellulose nitric membrane). The filtrate was analysed for major elements (Al, As, Ba, Ca, Cd, Cu, K, Mg, Mn, Pb and Zn) using inductively coupled plasma optical emission spectroscopy (ICP-OES). Results are also compared with leachates of pure compost to ensure observed metal concentrations originate from soil bound metals and not from the composts tested.

2.3 Ecotoxological experiments

Only LimeX70, Spent Mushroom and Green Waste of the five original composts were used.

2.3.1 Growth test using Greek Cress

The seedling growth and emergence test was conducted using Greek Cress (*Lepidium sativum*) and followed guidelines set out by the OECD [4]. A soil compost mixture of 5:1 was used, with 4 replicates for each mixture. 50 g of soil/compost were placed in a 10 cm diameter pot; to allow a higher retention of water, perlite was placed in the base of each pot. Sixty seeds were added to the surface of the soil in each sample pot. The surface area of the pot was divided into four quarters. 50 seeds were added to three quarters of the soil surface and the last 10 seeds were added to the remaining quarter to be used as a test for survival and germination. The germination test was conducted over a 48 hour period within the watered soils. After 14 days the cress shoots were harvested. The *L. sativum* stems and leaves of each sample were weighed, dried, ground to a fine powder and digested using nitric acid. Five cm^3 of nitric acid were used to digest each sample over 9 hours; 3 hours at 60°C and the last 6 hours at 110°C. The resulting solutions were then analysed for Al, As, Ba, Cd, Cr, Cu, Pb and Zn using ICP-OES.

2.3.2 Earthworm survival and condition test

The earthworm survival experiments used a standard earthworm *Eisenia fetida* and followed protocols modified from the OECD [4]. 500 g of soil/compost (in a mass ratio of 5:1) was weighed into acid washed plastic containers. The soil and compost mixture was made up to 60% of the water holding capacity, determined by submerging a known amount of soil (or soil/compost) in water over night, allowing the mixture to drain for 24 hours and then determining the moisture content by loss of mass after drying at 105 °C. Ten clitellate *E. fetida* were added to each container, with 0.5 g moist of horse manure. After 7 days the worms were counted to determine mortality. Surviving specimens were assessed using a condition index (C.I.), in which scores of 0, 1 and 2 are assigned to earthworms which are dead, moribund (flaccid and/or unresponsive to tactile stimulation) and responsive (alive and responsive to tactile stimulation) respectively [5].

2.3.3 ROTAS test

The ROTAS test was provided by Cybersense Biosystems Ltd, Oxford, UK. *Vibrio fischeri* bacteria were rehydrated for 50 minutes using an aqueous salt (NaCl) solution. 1 cm^3 of the bacteria suspension was added to each of 24 wells within a test plate. Next 1 cm^3 of each sample was added to a well; to each plate a blank (deionised water) and a control of a toxic Cu solution was included. After 15 minutes the percentage of luminescence emitted by the *V. fischeri* was measured. 100 to 71% luminescence indicated the sample was not toxic, 70 - 51% indicated the sample was moderately toxic and 50 – 0% indicated the sample was highly toxic [6].

2.4 Statistical analysis

Statistical analysis of the data produced was achieved by an untreated balance structure analysis of variance (ANOVA) test using GenStat [7].

Table 1: Chemical and physical conditions of the used soils.

Characteristic	Avonmouth	Devon Great Consols	Thamesmead
Available Phosphorus*	9.0	16.6	6.4
Available Potassium*	147	29	112
Available Magnesium*	91	12	53
Sand (%)	16	94	96
Silt (%)	50	3	3
Clay (%)	34	3	1
Textural Class	Silty Clay Loam	Sand	Sand
pH	6.6	4.1	7.3
Total Cu*	1274	1641	137
Total Zn*	9414	47.2	339
Total Pb*	16300	189	323
Total As*	191	34470	12.3
Total Cd*	239	813	2.35

* mg kg^{-1}

3 Results

The used composts and soils were analysed for nutrient and metal content (Table 1 and 2). The AM soil was highly contaminated with Pb, Zn and Cu, the DGC soil was highly contaminated with As, and the Thamesmead soil contained elevated levels of Cu, Zn and Pb. Amendment of the composts to the soils caused the pH of the leachates to be between 6.0 and 7.5. A leachate pH of around 4 was observed only for unamended DGC soil.

The results showed that there was a large variability in leaching effects between different composts. Figure 1 shows the leaching results for Zn from the three compost amended soils. All compost amendments were significantly different from the control ($p<0.05$). While GWC, LX and CC reduced the amount of Zn leaching from AM soil, SC and SMC increased leaching. Compost amendment to soil from DGC increased the leaching of Zn for SC and SMC ($p<0.05$) while the others were not significantly different from the control. Compost amendment to TM soil showed higher leachate concentrations after amendment of GWC, SMC and CC ($p<0.05$), but SC and LX were not significantly different from the control. The increased leaching caused by GWC, SMC and CC can be subscribed to Zn sources in these composts and not from the TM soil.

Leaching of Cu from AM soil (Table 3) was significantly higher after amendment of GWC, LX, SMC and SC. Similarly, leaching of Cu from TM soil increased for all tested composts but, apart from LX amendment, the increased leaching could be accounted for by Cu sources in the compost and not from the soil itself. In contrast, the leachate concentration of Cu in leachates from DGC soil was significantly reduced by the amendment of all composts.

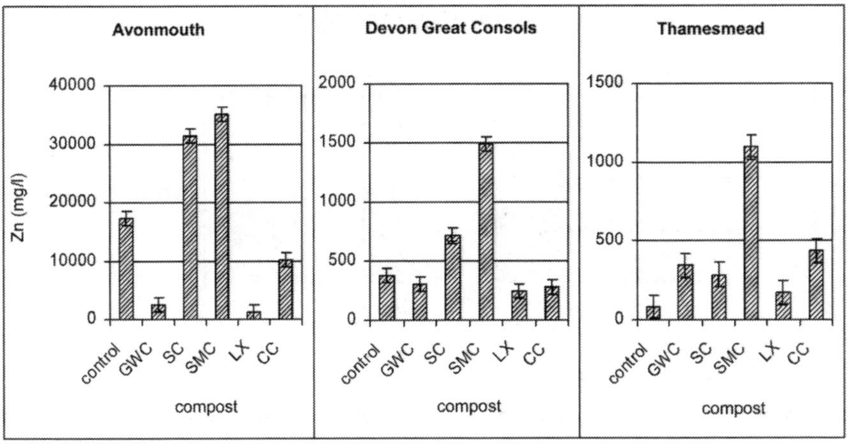

Figure 1: Leaching characteristics of Zn leaching from metal contaminated soils amended with compost. GWC = Green Waste Compost, SC = Sewage sludge Compost, SMC = Spent Mushroom Compost, LX = Limex70, CC = coir compost. Error bars are standard error of the means.

Table 2: Leachate concentrations (mg l^{-1}) of Cu and Pb from compost amended soils.

	Avonmouth		Devon Great Consols		Thamesmead	
	Cu	Pb	Cu	Pb	Cu	Pb
se.	72.4	82.2	149.5	19.8	4.0	7.98
Control	89	242	1078	86.6	2.2	39.3
CC	280	275	43**	24.4**	51.2**	40.0
GWC	382**	251	42**	21.5**	46.0**	36.4
LX	1140**	249	5**	12.9**	48.2**	16.8
SMC	2180**	936**	274**	111.7	204.4**	42.8
SC	559**	121	56**	11.4**	29.0**	13.6

se. = standard error of the means, degree of freedom = 15.
**Significant different from the control ($p<0.05$)

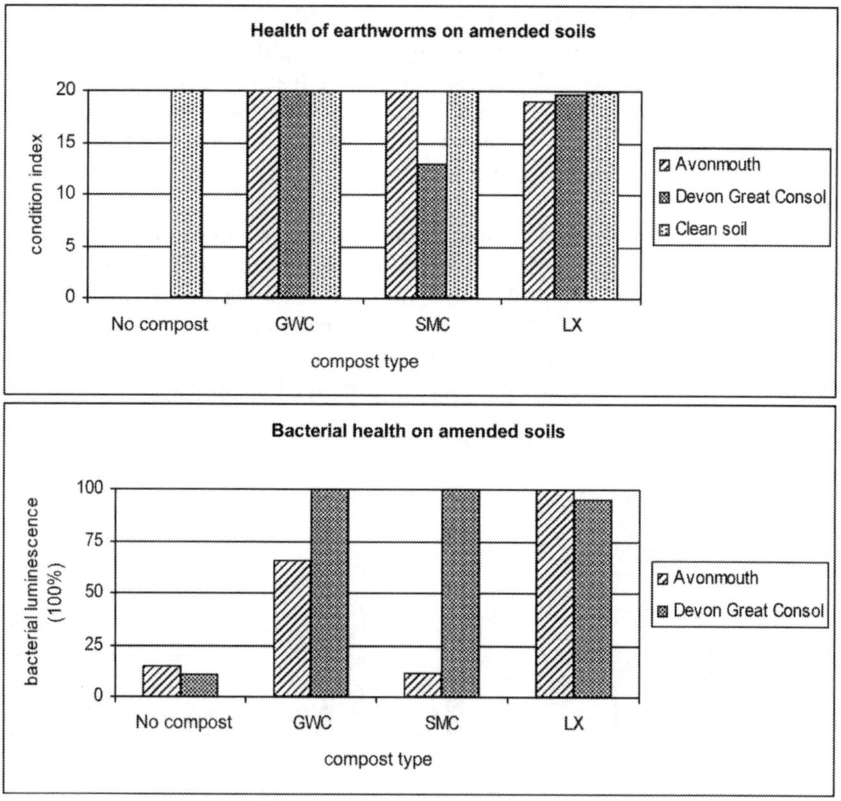

Figure 2: Toxicity of Avonmouth and Devon Great Consols soils amended with different composts. The toxicity was examined with earth worms (*Eisenia fetida*) (20 = not toxic) and the bacterial strain *Vibria fischeri* (100 = not toxic). GWC = Green Waste Compost, SMC = Spent Mushroom Compost, LX = Limex70. Error bars are standard error of the means.

The leaching of Pb from AM soil (Table 3) was only increased by SMC and no reduction could be observed by other composts. For the DGC soil, CC, GWC, LX and SC reduced the leaching of Pb. For TM soil no effect was observed for the leaching of Pb by any compost.

Uptake of Cu, Pb and Zn into Greek cress growing on soils amended with GWC, SMC or LX are given in Table 4. For the AM soil, all composts reduced the uptake of Cu, Pb and Zn compared to the control. On the DGC soil no cress grew on unamended soil so no comparison with the control could be made. Nevertheless for all metals LX seemed to cause the highest uptake compared to GWC and SMC. For the TM soil, different trends could be observed for the different metals. Uptake of Cu was increased by LX, Pb uptake was reduced by GWC and uptake of Zn was reduced by GWC and LX but increased by SMC.

Results of ecological tests with bacteria and earthworms are given in Figure 2. The tests with earthworms gave a significant difference between treated and untreated soils for all composts. In most cases the condition of the worms was restored to that of uncontaminated soil. Only SMC amendment to DGC soil failed to restore the toxicity to within 95% of the clean soil. The results of the bacterial tests (Figure 2) showed that only LX completely reduced the toxicity of AM soil, GWC only partly reduced the toxicity and SMC had no effect at all. The toxicity of the DGC soils for the bacteria was completely reduced by GWC and SMC and almost completely by LX.

Table 3: Metal uptake in Greek cress growing on metal contaminated soil amended with composts. GWC = Green Waste Compost, LX = Lymex70 and SMC= Spent Mushroom Compost.

	se.	control	GWC	LX	SMC
Cu					
Avonmouth	40.0	247	43*	56*	56*
Devon Great Consols	25.2	-	42	135**	59
Thamesmead	12.8	28.8	22.7	76.1*	48.5
Pb					
Avonmouth	150.1	1625	247*	254*	272*
Devon Great Consols	36.6	-	179**	383**	92
Thamesmead	13.9	106.0	49.7*	125.7	108.4
Zn					
Avonmouth	175.7	3708	1218*	623*	869*
Devon Great Consols	25.8	-	153**	171**	93**
Thamesmead	7.12	190.9	164.0*	158.6*	281.7*

se. = standard error of means
* significant different from control ($p<0.05$)
** significant different from zero where no control ($p<0.05$).

4 Discussion

The results presented in this paper differ between the three soils and also between the composts. The leaching results show that GWC, LX and CC in general reduce the leaching of metals from the soils. Only in the case of AM soil is the leaching of Cu from the soil increased by GWC and LX. SMC and SC in general seem to increase the leaching of soil metals but for these composts contradictory effects were also observed, such as reduced leaching of Cu by amendment of SMC and SC to the DGC soil. This means that all tested composts have the potential to reduce as well as increase the risk to potential receptors at contaminated sites remediated by compost amendment. Observations of increased leaching are rarely mentioned in literature and usually only mentioned in case of As contamination [2].

The trends observed for metal uptake by Greek cress do not always follow the same trends as observed in the leaching tests. For example, SC generally increased leaching and might indicate a higher bioavailability, but uptake of Cu, Pb and Zn on AM soil was reduced by this compost. However, the GWC, LX and SMC all reduced metal uptake although with some exceptions. Our results indicate that a specific compost may reduce uptake of one metal but increase it for an other, and these effects can be different for different kinds of soil. An explanation for the variety in uptake between metals, composts and soils could be the formation of metal complexes with organic matter from the composts [8], as reported for a mixture of organic household waste compost and wood chips where the concentration of free Cu^{2+} in the soil water was reduced but the concentration of complexed Cu was increased [9]. Depending on the type of complex, these complexes could be taken up and transported within the plant. However, only one plant type has been tested in our study and further studies with a variety of plants are recommended to incorporate differences in physiology between plants.

Despite the increased metal leaching and in some cases increased uptake, GWC, SMC and LX efficiently reduced the toxicity to earthworms on AM and DGC soils. For bacteria the same could be observed with exception of SMC that had no effect on AM soil. The improved conditions of the earthworms can be partially ascribed to improved pH conditions (especially in the case of DGC soil) since the earthworms used prefer a pH between 5.5 and 8 [10]. The major factor increasing earthworm health is likely to be the increased amount of soil organic matter caused by the addition of compost. Earthworm population size is generally related to soil organic matter content [11]. Compost therefore improves the soil conditions for the worms but the earthworm test results might therefore indicate little about the actual toxicity of the soil mixtures for the worms. Nevertheless the higher performance of the bacteria in compost treated soils in most cases suggests good possibilities to use the tested composts to reduce the toxicity of metal contaminated soils.

5 Conclusions

In general, all the tested composts have the ability to improve the soil conditions of a contaminated site. However, negative effects can also be obtained because there is a variety of effects on metal mobility, uptake and toxicity between composts and also between soils they are applied to. Care should therefore be taken before applying the materials at operational scale in the field. The results suggest that each time a compost is proposed for remediation of metal contaminated soils it should be tested with the proposed soil in leaching tests, plant tests and toxicity tests to reduce risk of toxic metals reaching potential receptors in the environment. Our results also suggest that more rigorous tests should be undertaken to understand the mechanisms involved in metal complexation using different compost types, in order to optimise the use of compost as a remediation technique.

Acknowledgements

The authors are grateful to the following people who have generously supplied materials for the research or assisted with the practical work: Adrian Nash from Halcrow who provided the contaminated soil and also arranged the transport; Andy Johnson and Andrew Weeks from Thames Water - Terra Eco Systems for supplying the composted sewage sludge. The financial contribution from the EPSRC, who funded part of this work under the SUBR:IM consortium (Grant GR/S148809/01) and the NERC MSc studentship for Tim Laverye were grateful acknowledged.

References

[1] Environment-Agency, The state of soils in England and Wales. Environment Agency, Bristol. 2004.

[2] Gadepalle, V. P., Ouki, S. K., van Herwijnen, R. & Hutchings, T., Immobilisation of heavy metals in soil using natural and waste materials for vegetation establishment on contaminated sites. Soil Sed. Contam., pp. in press.

[3] BSI, Characterisation of waste - Leaching - Compliance test for leaching of granular waste materials and sludges. BS En 12457: Part 2: One stage batch test at a liquid to solid ratio of 10 l/kg for materials with particle size below 4 mm (without or with size reduction). British Standards Institution, London. 2002.

[4] OECD, OECD Guideline for testing of chemicals -208- "Terrestrial Plants, Growth Test", vol. 208. OECD - Organisation for Economic Co-operation and Development, Paris. 1984.

[5] Langdon, C. J., Piearce, T. G., Meharg, A. A. & Semple, K. T., Survival and Behaviour of the Earthworms *Lumbricus rubellus* and *Dendrodrilus rubidus* from Arsenate-Contaminated and Non-Contaminated Sites. Soil Biol. Biochem. **33**, pp. 1239-1244. 2001.

[6] Loureiro, S., Ferreira, A. L. G., Soares, A. M. M. & Nogueira, A. J. A., Evaluation of the toxicity of two soils from Jales Mine (Portugal) using aquatic bioassays. Chemosphere **61**, pp. 168-177. 2005.

[7] Genstat, The Guide to GenStat Release 8.1 Part 2: Statistics. *In* R. W. Payne (ed.), 8.1 ed. Lawes Agricultural Trust (Rothamsted Experimental Station) - VSN International, Oxford, 2005.

[8] da Silva, J. & Oliveira, C. J. S., Metal ion complexation properties of fulvic acids extracted from composted sewage sludge as compared to a soil fulvic acid. Water Res. **36**, pp. 3404-3409. 2002.

[9] Kiikkila, O., Perkiomaki, J., Barnette, M., Derome, J., Pennanen, T., Tulisalo, E. & Fritze, H., In situ bioremediation through mulching of soil polluted by a copper-nickel smelter. J. Environ. Qual. **30**, pp. 1134-1143. 2001.

[10] Kaplan, D. L., Hartenstein, R., Neuhauser, E. F. & Malecki, M. R., Physiochemical Requirements in the Environment of the Earthworm *Eisenia foetida*. Soil Biol. Biochem. **12**, pp. 347-352. 1980.

[11] Edwards, C. A. & Bohlen, P. J., Biology and Ecology of Earthworms, 3rd ed. Chapman and Hall, London. 1996.

Effective gas-phase diffusion coefficient in soils

C.-S. Fen
Department of Environmental Engineering and Science, Feng Chia University, Taichung, Taiwan, Republic of China

Abstract

Determination of the effective gas-phase diffusion coefficient is important to quantify gas diffusion for a number of environmental pollution problems involving migration and release of hazardous gases or organic vapors in soils. Conventionally, the only considered transport mechanism for this coefficient is molecular diffusion. This study additionally considers the effects of Knudsen diffusion and nonequimolar flux on this coefficient and proposes a relationship of this coefficient with both molecular and Knudsen diffusion coefficients, ratio of molar masses and composition of component gases for gas diffusion in soils. Through investigating calculation results of this relationship for hypothetical diffusion systems having different soil permeabilities and diffusing gases at a range of water saturation levels, it is found that the Knudsen and nonequimolar effects can only be substantial for systems with dry soil permeabilities less than 10^{-13} m^2 at any saturation. The Knudsen effect on this relationship is further demonstrated through comparing results obtained from this relationship to the measured effective diffusion coefficients for 72 soil cores, which all have high silt and clay contents, given in the literature.
Keywords: gas diffusion in soil, molecular diffusion, Knudsen diffusion, diffusibility.

1 Introduction

Volatile organic compounds (VOCs) are common groundwater contaminants threatening our environment. Volatile contaminants can quickly spread in the subsurface by vaporization at dissolved or free phase and by gas phase transport through soil pores. Gas diffusion is an important transport mechanism controlling migration and distribution of volatile contaminants. Other environmental pollution problems such as emissions of methane, CO_2 or organic vapors from landfills [20], emissions of fumigant and pesticide [14] and migration of naturally generated radon through soils and foundation materials [16] can also be dominated by gas diffusion. Besides, mass of volatile chemicals can naturally be reduced through gas diffusion in contaminated soils. Therefore, an accurate quantification of gas diffusion becomes important for assessment of natural attenuation, remediation and monitoring network design and risk assessment of contaminated sites.

Fick's first law of diffusion is conventionally applied to compute diffusive flux of a gas component in soils. In the application, the air-filled gas-phase diffusion coefficient is modified by a diffusibility factor, leading to the so-called effective diffusion coefficient, to account for the impedance to diffusion caused by the tortuous nature of pores in soils. Determination of the diffusibility factor is important in predicting the magnitude of gas diffusion, and thereby delineating other transport and fate processes in soils.

In the past, numerous experimental studies proposed various relationships of the diffusibility factor with soil porosity, air-filled porosity and some empirical constants, e.g., [5, 8, 22, 26]. Millington and Quirk [18], Ball [4], Friejer [10] and Schaefer et al. [24] treated the porous structure of soil as joint tubes and developed theoretical models for this factor. Van Duin [27] and Lahvis et al. [17], however, used experimental and/or field data to determine the diffusibility factor as functions of air-filled porosity and several fitting constants. Moldrup et al. [19] and Weerts et al. [30] assessed the dependence of this factor on soil water characteristic parameters of disturbed and undisturbed soils.

A rigorously constitutive relationship of component gas diffusive flux in soils is described by the Dusty Gas Model (DGM) equations [7]. The component diffusive flux for multicomponent gas diffusion in soils can be a Fickian expression only under specific conditions. For example, Jaynes and Rogowski [12], Amali et al. [2] and Webb and Pruess [29] mentioned a Fickian expression is valid only for a binary gas diffusion system with a trace gas and for a ternary gas diffusion system with a stagnant gas. Fen and Abriola [9] used the DGM equations to obtain the component diffusive flux in a Fickian form for a two-component gas transport system with gas components of equal molecular masses under isothermal and isobaric conditions. A correction factor, called the obstruction factor, is defined in the DGM, to account for the obstruction of soil particles to molecular diffusion. Brugeman [6], Neale and Nader [21] and Abu-El-Sha'r and Abriola [1] applied the DGM theory and conducted equimolar countercurrent diffusion experiments in porous systems to determine this factor as functions of air-filled porosity. Abu-El-Sha'r and Abriola [1] also determined the Knudsen diffusion coefficient for single gas flow systems with the same soil samples as the ones used for determining the obstruction factor in two-component gas diffusion experiments. However, the Knudsen effect is not pronounced in their diffusion experiments, as a result that the diffusibility factor is equivalent to the obstruction factor.

It is shown that Fick's law is generally applied to gas diffusion in soils. A lot of experiments have been done to find the relationships among the effective diffusion coefficient, air-filled molecular diffusion coefficient, total and air-filled porosities of soil. Seldom has considered the coupling effect of molecular and Knudsen diffusions in low permeability soil systems and the nonequimolar effect due to unequal molecular masses of the diffusing and ambient gases on gas diffusion. For the application of Fick's law to the interpretation of laboratory data for gas diffusion in soil columns, two questions need to be clarified: (1) what is the specific soil condition under which Knudsen diffusion is substantial and should be considered in the effective diffusion coefficient? (2) How significant is the nonequimolar effect on gas diffusion in soil systems?

This study intends to resolve these questions by conducting a series of laboratory experiments and assessing the dependence of the diffusibility factor on the Knudsen diffusion coefficient, gas component and composition. These experiments are being undertaken and aim at independently determining the obstruction factor for two-component gas diffusion and the permeability and Knudsen diffusion coefficients for single gas flow in low permeability soil systems with the same soil samples. A literature survey of existing correlations of the effective diffusion coefficient was also carried out and the corresponding experimental data was used to analysis the relative importance of molecular and Knudsen diffusions and the nonequimolar effect. This paper presents this study result.

2 Materials and methods

Without considering surface, temperature diffusions and the gravitational effect, the DGM equations are expressed as

$$\sum_{j=1, j\neq i}^{n} \frac{X_i N_j^D - X_j N_i^D}{Q_m D_{ij}^o} - \frac{N_i^D}{D_i^K} = \frac{1}{RT} \nabla p_i \qquad i = 1,2,....,n \tag{1}$$

in which n is the total number of gas components in the gas mixture; X_i is the mole fraction of component i; $\hat{N}_i^D$ is the total diffusive flux of component i [mol m^{-2}s^{-1}]; T is the absolute temperature [^{o}K]; R is the universal gas constant [kg m^{2}sec^{-2} ^{o}K^{-1} mol^{-1}]; p_i is the partial pressure of component i [kg m^{-3} s^{-2}]; D_{ij}^o is the air-filled binary molecular diffusion coefficient of gas pair i and j [m^{2}sec^{-1}]; Q_m is the obstruction factor; and D_i^K is the Knudsen diffusion coefficient of component i [m^{2}sec^{-1}]. Thorstenson and Pollock [25] provided the following relationship for D_i^K,

$$D_i^K = \frac{k_e b_i}{\mu_i} \tag{2}$$

where k_e is the gas effective permeability [m^2] and $k_e = kk_{rg}$ in which k is the intrinsic permeability [m^2] and k_{rg} is the gas relative permeability; μ_i is the dynamic viscosity of component i [kg m^{-1}s^{-1}]; b_i is the Klinkenberg parameter [Pa]. According to [23], b_i can be determined by the following relationship

$$b_i = 5.57 k_e^{-0.24} \frac{\mu_i}{\mu_{air}} \sqrt{\frac{M_{air}}{M_i}} \tag{3}$$

where M_{air} and M_i are molar masses of air and component i [kg mol^{-1}].

According to [9], for a binary system of components A and B under isobaric condition, the total diffusive flux for each component can be written explicitly by manipulation of eqn (1):

$$N_A^D = -\frac{Q_m D_{AB}^o D_A^K C}{Q_m D_{AB}^o + D_A^K (1 - a_A X_A)} \nabla X_A \tag{4}$$

where $C = \sum C_i$ is the molar concentration of the gas mixture (mol m^{-3}); $a_A = 1 - (M_A / M_B)^{1/2}$.

When both components have equal molecular masses in the system, the flux is expressed as

$$N_A^D = -\left(\frac{1}{D_A^K} + \frac{1}{Q_m D_{AB}^o} \right)^{-1} C \nabla X_A \tag{5}$$

Fick's law of diffusion applied to gas diffusion in soil systems is expressed as

$$N_i^D = -Q D_{ij}^o \nabla C_i \tag{6}$$

where Q is the diffusibility factor and is generally used to account for the impedance of soil particle to gas diffusion. A comparison of eqns (4) and (6) reveals that the diffusibility factor defined in Fick's law of diffusion can be composition-dependent and is related to the obstruction factor as follows:

$$Q = \frac{Q_m D_A^K}{Q_m D_{AB}^0 + D_A^K (1 - a_A X_A)} \tag{7}$$

Eqn (7) can be further simplified to become: (1) $Q = \frac{Q_m D_A^K}{Q_m D_{AB}^0 + D_A^K}$, for the system with equal molecular mass components ($\boldsymbol{a}_A = 0$) or involving diffusion of a dilute gas in any gas mixture ($\boldsymbol{X}_A \approx 0$); (2) $Q = \frac{D_A^K}{D_{AB}^0}$, for gas transport dominated by Knudsen diffusion ($\boldsymbol{D}_A^K << \boldsymbol{D}_{AB}^o$); (3) $\boldsymbol{Q} = \boldsymbol{Q}_m$, for gas transport dominated by molecular diffusion ($\boldsymbol{D}_A^K >> \boldsymbol{D}_{AB}^o$).

The relationship shown in eqn (7) reveals that independent experiments for assessing both the obstruction factor and the Knudsen diffusion coefficient for the same soil sample are required to verify the dependence of the diffusibility factor on these two parameters. Obviously, these experiments need to be conducted in low permeability soil columns and different diffusing gases should be considered to investigate the nonequimolar effect.

To identify the relative importance of molecular and Knudsen diffusions under a variety of soil conditions, the diffusibility is assessed for a series of hypothetical gas diffusion systems. The diffusibility factor is calculated according to eqn (7) with neglecting the X_A term (i.e. assuming diffusing gas is dilute in these hypothetical systems) and compared to the obstruction factor, which is calculated based on Millington and Quirk relationship. Soil systems with different permeabilities and diffusing gases at different levels of water saturation were considered in the calculation. Note that the Knudsen diffusion coefficient is related to gas permeability, which depends on soil type and water saturation, i.e., eqns (2) and (3).

In this study, the measured results of the diffusibility factor for different gas diffusion experiments, specifically, in low permeability soils, were collected from literature, which include those presented in [3, 4, 10, 11, 13, 15, 28, 31], for totally 72 soil cores, which all have sand content less than 15 %. Different gas components were used as a diffusing gas in these experiments. Air is used as an ambient gas. For most of these diffusion experiments, soil or gas permeability was not measured. For assessing the Knudsen effect on gas diffusion in these soil systems, an empirical relationship of soil permeability with soil composition is used in this study

$$k = \frac{1}{m}\left[\frac{(1-\theta_T)^2}{\theta_T^{\ 3}}\left(\frac{\alpha}{100}\sum\frac{f_i}{d_m}\right)^2\right]^{-1} \tag{8}$$

where θ_T is the total porosity of the soil sample; m is a packing factor determined experimentally to be about 5; α is a sand-shape factor varying from 6 for spherical grains to 7.7 for angular grains; f_i is the fraction of sand held between adjacent sieves; and d_m is the geometric mean of the rated size of adjacent sieves. Note that eqn (8) does not work well for the porous medium composed of fine-grained material.

3 Results and discussion

The diffusibility factor was calculated based on eqns (2), (3) and (7) for several hypothetical diffusion systems, in which CO, TCE or CH_4 is considered as a diffusing gas and N_2 is an ambient gas. This calculation also considers a range of soil permeabilities from 10^{-10} to 10^{-16} m^2 and water saturation levels between 0.4 and 0.8. The obstruction factor was calculated based on [18]. A reduction of gas permeability due to partial saturation of soil is evaluated according to Brook and Corey's relationship. Figure 2 presents the percentage difference between the calculated values of the diffusibility and the obstruction factors related to water saturation of soil and the corresponding soil effective permeability. The difference is minimal for the diffusion systems with dry soil permeabilities greater than 10^{-13} m^2. This implies the Knudsen effect can be neglected for gas diffusion in soil systems within this dry soil permeability and the tested saturation ranges. For soil systems with dry soil permeabilities less than 10^{-13} m^2, the difference starts to be significant, which is 6% at least and 75% at most for these studied cases. Soils in this permeability range have high silt and clay contents. Besides, the diffusibility values are less than those of the obstruction factor when considering the Knudsen effect in the diffusibility. Different diffusing gas also makes the difference changed even though the gas is dilute in the system. For all of the studied cases, the difference for the CO/N_2 system is the greatest and for the CH_4/N_2 system is the least. As dry soil permeability decreases, discrepancies of the difference between different diffusing gas systems also increase. Note that the difference does decrease as the soil water saturation increases.

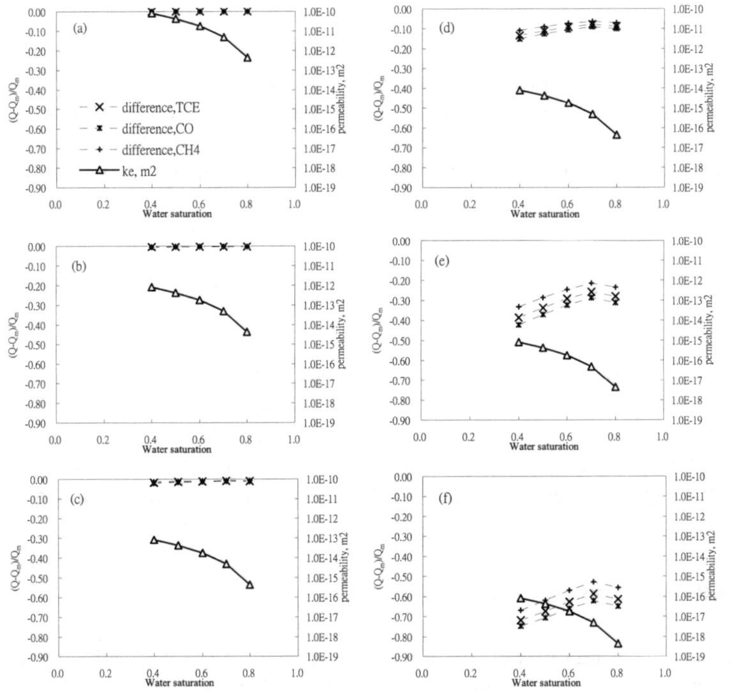

Figure 1: Comparison of values of the diffusibility and the obstruction factors for gas diffusion systems with a dry soil permeability of: (a) 10^{-10} m^2; (b) 10^{-12} m^2; (c) 10^{-13} m^2; (d) 10^{-14} m^2; (e) 10^{-15} m^2; (f) 10^{-16} m^2, at different water saturation levels

Eqn (7) is also used to evaluate the diffusibility factor based on the experimental data given in articles by assuming that the corresponding diffusion experiments all involve a dilute diffusing gas. Results were compared to the diffusibility values provided in the articles as shown in figure 2. Values of the obstruction factor in eqn (7) for [4, 28, 31] were calculated based on the correlation provided in [1] and for the others are based on [18]. Soil permeability values for the tested soil cores in these experiments were not provided in the referred articles. Hence, they were determined by fitting the diffusibility values obtained from eqn (7) to the ones provided in these articles. The permeability values obtained by the fitting are in the range between 10^{-13} and 10^{-20} m^2, which is consistent with the permeability range for soils having high silt and clay contents. In this permeability range the difference between the diffusibility and obstruction factors is substantial as shown in figure 1. This implies that soil permeability (or the Knudsen diffusion coefficient) is an important factor affecting gas diffusion in these experiments and should be considered in the diffusibility factor. Permeabilities of the soil cores in these diffusion experiments are also evaluated according to eqn (8) and compared to the permeability values obtained by the fitting. Figure 3 shows that the permeability values obtained with the empirical equation are generally greater than the ones obtained by the fitting, implying that soil permeability is overpredicted by eqn (8).

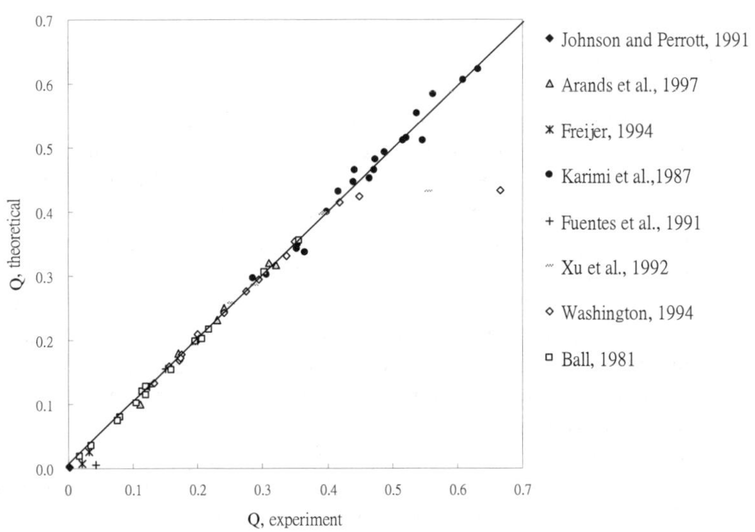

Figure 2: Comparison of the diffusibility factors calculated with eqn (7) and those given in the articles.

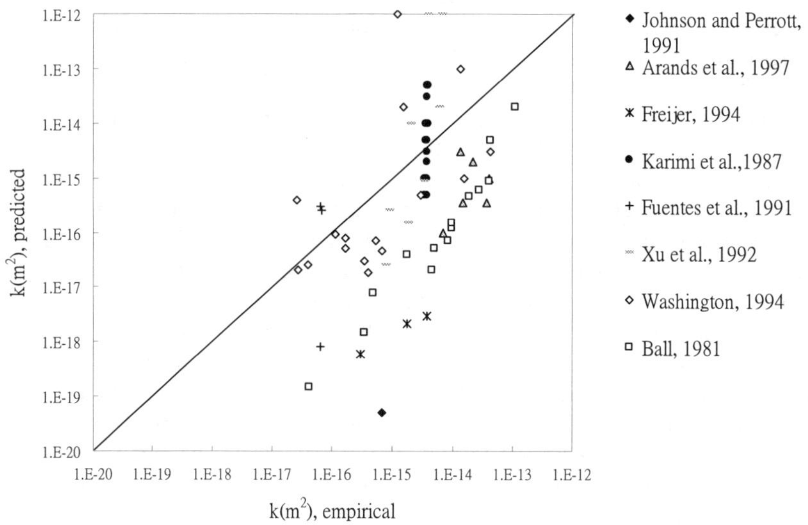

Figure 3: Comparison of the permeability values determined from fitting the diffusibility data and the ones calculated with eqn (8).

4 Conclusions

According to the study result, some conclusions are made as follows:

(1) The diffusibility factor for Fick's law of diffusion is identical to the obstruction factor defined in the DGM equations for diffusion systems with dry soil permeabilities greater than 10^{-13} m^2 at water saturation levels ranging from 0.4 to 0.8. However, for systems with lower dry soil permeabilities at any saturation, these two factors are not identical.

(2) Use of different gas components in diffusion experiments can obtain different diffusibility results for systems with dry soil permeabilities less than 10^{-13} m^2 even though diffusing gases are dilute in the systems.

(3) Incorporation of the Knudsen diffusion coefficient into the diffusibility factor helps to adequately assess gas diffusion in soil systems. For this assessment, the obstruction factor and soil permeability should be determined independently through two-component gas diffusion and single gas flow experiments on the same soil sample. The Knudsen effect on the diffusibility factor is demonstrated through comparing results obtained with this diffusibility to the measured diffusibilities for 72 soil cores, which all have high silt and clay contents, given in papers.

Acknowledgements

Funding of this research was provided by the National Science Council in Taiwan.

References

[1] Abu-El-Sha'r, W. & Abriola, L.M., Experimental assessment of gas transport Mechanisms in natural porous media: parameter evaluation. *Water Resources Research*. 33(4), pp. 505-516, 1997.

[2] Amali, S., Rolston, D. E. & Yamaguchi, T., Transient multicomponent gas-phase transport of volatile organic chemicals in porous media. *J. Environ. Qual.*, 25, pp. 1041-1047, 1996.

[3] Arands, R.; Lam, T., Massry, I., Berler, D.H., Muzzio, F. J. & Kosson, D.S., Modeling and experimental validation of volatile organic contaminant diffusion through an unsaturated soil. *Water Resources Research* 33(4), pp. 599-609, 1997.

[4] Ball, B. C., Modeling of soil pores as tubes using gas permeabilities, gas diffusivities and water release. *Journal of Soil Science*, 32, pp. 465-481, 1981.

[5] Bartelt-Hunt, A. L. & Smith, J. A., Measurement of effective air diffusion coefficients for trichloroethene in undisturbed soil cores. *Journal of Contaminant Hydrology*, 56, pp. 193-208, 2002.

[6] Brugeman, D.A.G., *Ann. Phys.*, 24, 636, 1935.

[7] Cunningham, R.E. & Williams, R.J., *Diffusion in gases and porous media*, New York: Plenum, 1980.

[8] Currie, J. A., Movement of gases in soil respiration. *Topic in Sorption and Transport Processes in Soils*, 37, *Monogr. Soc. Chem. Ind.*, pp. 152-171, Rothamsted Exp. Stn., Harpenden, England, 1970.

[9] Fen, C.-S. & Abriola, L.M., A comparison of mathematical model formulations for organic vapor transport in porous media. *Advances in Water Resources*, 27(10), pp.1005-1016, 2004.

[10] Freijer, J. I., Calibration of jointed tube model for the gas diffusion coefficient in soils. *Soil Sci. Soc. Am. J.*, 58, pp. 1067-1076, 1994.

[11] Fuentes, H. R., Polzer, W. L. & Smith, J. L., Laboratory measurements of diffusion coefficients for trichloroethylene and orthoxylene in undisturbed tuff, *J. Environ. Qual.*, 20, pp. 215-221, 1991.

[12] Jaynes, D.B. & Rogowski, A.S., Applicability of Fick's law to gas diffusion, *Soil Sci. Soc. Am. J.*, 47, pp.425-430, 1983.

[13] Johnson, R. L. & Perrott, M., Gasoline vapor transport through a high-water-content soil. *Journal of Contaminant Hydrology*, 8, pp.317-334, 1991.

[14] Jury, W.A., Russo, D., Streile, G. & Abd, A.E., Evaluation of volatilization by organic chemicals residing below the soil surface. *Water Resour. Res.*, 26(1), pp. 13-20, 1990.

[15] Karimi, A. A., Farmer, W. J. & Cliath, M. M., Vapor-phase diffusion of benzene in soil. *J. Environ. Qual.*, 16(1), pp. 38-43, 1987.

[16] Kozak, J.A., Reeves, H.W., Lewis, B. A., Modeling radium and radon transport through soil and vegetation. *J. of Con. Hyd.*, 66, pp. 179-200, 2003.

[17] Lahvis, M.A., Baehr, A.L. & Baker, R.J., Evaluation of volatilization as a natural attenuation pathway for MTBE, *Ground Water*, 42(2), pp. 258-266, 2004.

[18] Millington, R. J. & Quirk, J. P., Permeability of porous media. *Trans. Faraday Soc.*, 57, pp.1200-1207, 1961.

[19] Moldrup, P., Olesen, T., Schjonning, P., Yamaguchi, T. & Rolston, D.E., Predicting the gas diffusion coefficient in undisturbed soil from soil water characteristics. *Soil Sci. Soc. Am. J.*, 64, pp. 94-100, 2000.

[20] Nastev, M, Therrirn, R., Lefebvre, R. & Gelinas, O., Gas production and migration in landfills and geological materials. *J. of Conta. Hydro.*, 52, pp. 187-211, 2001.

[21] Neale, G. H. & Nader, W. K., *A. I.Ch.E. Journal*, 19,112, 1973

[22] Penman, H. L., Gas and vapour movement in the soil, I, the diffusion of vapour through porous solids. *J. Agric. Sci.*, 30, pp. 437-462, 1940.

[23] Reinecke, S.A. & Sleep, B.E., Knudsen diffusion, gas permeability, and water content in an unsaturated porous medium. *Water Resour. Res.*, 38(12), pp. 1280-1294, 2002.

[24] Schaefer, C. E., Arands, R. R., van der Sloot, H. A. & Kosson, D. S., Modeling of the gaseous diffusion coefficient through unsaturated soil systems. *Journal of Contaminant Hydrology*, 29, pp. 1-21, 1997.

[25] Thorstenson, D. C. & D. W. Pollock, Gas transport in unsaturated zones: Multicomponent systems and the adequacy of Fick's law. *Water Resources Research*, 25 (3), pp. 477-507, 1989.

[26] Troeh, F.R., Jabro, J. D. & Kirkham, D. *Geoderma*, 27, 239, 1982.

[27] Van Duin, R.H.A., Over De. Invloed Van Grondbwerking Op Het Transport Vvan Warmte. *Water En Lucht, Verst Landbouwk Orderz 62.7*, Staatsdrukkerij, The Hague, 81pp, 1956.

[28] Washington, J.W., Rose, A. W., Ciolkosz, E. J. & Dobos, R. R., Gaseous diffusion and permeability in four soil profiles in central Pennsylvania, *Soil Science*, 157(2), pp. 65-76,1994.

[29] Webb, W. S. & Pruess, K., The use of Fick's law for modeling trace gas diffusion in porous media. *Transport in Porous Media*, 51, pp. 327-341, 2003.

[30] Weers, A.H., Freijer, J.I. & Bouten, W., Modeling the gas diffusion coefficient in analogy to electrical conductivity using a capillary model. *Soil Sci. Soc. Am. J.* 64, pp. 527-532, 2000.

[31] Xu, X., Nieber, J.L. & Gupta, S.C., Compaction effect on the gas diffusion coefficient in soils. *Soil Sci. Soc. Am. J.* 56, pp. 1743-1750, 1992.

Section 5
Case studies

Remediation and development of a former military manufacturing and test site in Florida USA

R. L. Olsen[1], A. B. Woods[2] & M. H. Edgar[3]
[1]*CDM, Denver, Colorado USA*
[2]*CDM, Orlando, Florida USA*
[3]*CDM, Tallahassee, Florida USA*

Abstract

Approaches for the redevelopment of contaminated properties vary depending upon many factors including type/extent of contamination, risks, regulatory authorities and compliance, remediation cost, the resources/desires of the owners, compliance with Federal SEC reporting and the value/demand for the property. Reducing the remediation cost and cleanup time of groundwater are common objectives at many sites because groundwater remediation is typically the most expensive, time consuming and difficult media to cleanup. Innovative techniques to reduce groundwater remediation timeframes at the Universal City Property Management site in Florida USA include the following:

- Use of horizontal wells to effectively capture and recover groundwater.
- Use of chemical oxidation to rapidly cleanup the contaminated groundwater.
- Use of horizontal wells to distribute chemical oxidants.
- Creation and use of open pit excavation to directly expose and aerate the resulting groundwater *in situ* and remove contamination.

Other approaches and elements that contributed to successful development of the site include:

- Flexible and dynamic remediation plans that respond to the owners need to sell or develop property to maintain a balanced cash flow.
- Effective and rapid remediation technologies (e.g., chemical oxidation) to facilitate property sale and development.
- Proactive coordination with regulatory agencies to build consensus and shorten review time of work plans and reports.
- Large area in a desirable location with great potential for development and revenue generation.

All of these approaches and attributes of the site contributed to the successful remediation and redevelopment.

Keywords: contaminated sites, cleanup, remediation, redevelopment.

www.witpress.com, ISSN 1743-3541 (on-line)
doi:10.2495/BF060121

1 Introduction

The investigation, remediation, and development of contaminated sites in the USA occurs in many different ways using various approaches depending upon the type/extent of contamination, the environmental and human health risk, the application of appropriate regulatory limits, the remediation cost, the owners/developers intended future use of the site, and the property value and demand. The remediation and development of a formerly operated defense manufacturing and weapons systems test site in Orange County, Florida is discussed in the following sections. This site is not an official "Brownfields" site as designated by either Federal or State Agencies. However, the site is contaminated and is currently being remediated and developed into valuable commercial property using unique and innovative approaches. The site is very large in size and is located in highly desirable market area; therefore, the site has a great potential for extensive development and increase in value. For this site, the following items are discussed:

- Site background including types and sources of contamination
- Cleanup goals
- Remediation approaches
- Redevelopment status

2 Universal City Property Management South Campus Project

2.1 Site background

The project site is located in Orlando, Florida USA (see Figure 1). The site is approximately 2,000 acres in size and was formerly owned by a defense contractor. From 1958 to 1998, the defense contractor developed, manufactured, and tested weapon related equipment and systems. The site is currently surrounded by extensive development including the Orange County Convention Center, hotels and various tourist related attractions. In 1998, Universal City Property Management III LLC (UCPM) purchased the site from the previous owner and the site is now referred to as the UCPM South Campus. UCPM is a subsidiary of Stan Thomas Enterprises, a successful national commercial and industrial development company. Following UCPM's purchase, a comprehensive corrective action program was initiated to remediate soil and groundwater contamination attributed to past industrial use.

Nine main areas of the site have been identified for remediation (see Figure 1). Six of these areas are former landfills (Landfills 1 - 6) and three of the areas are former test areas (Launch Test Area, Pershing/Waste Conservation Area, and Ordnance 2 Area). The aerial extent of the six landfills was approximately 25.5 acres. Landfill material requiring remediation included electroplating sludge, waste solvents and miscellaneous debris. Contaminated waste/debris/soil at the former landfills requiring excavation and disposal was up to 19 below land surface (bls). Besides contaminated waste, debris and soil, each

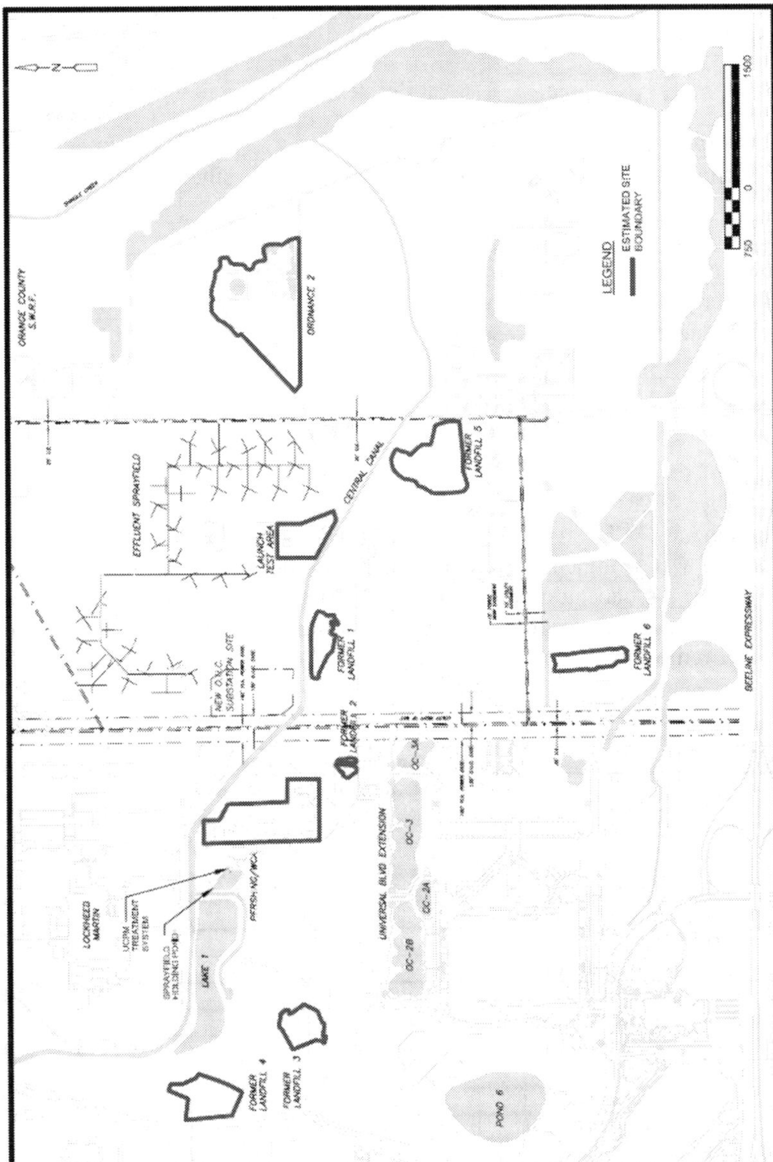

Figure 1: UPCM South Campus.

of the areas has associated groundwater contamination. Groundwater below the site exists in a series of interconnected and unconfined and semiconfined aquifer units including (top to bottom):

Table 1: Aquifer descriptions.

Aquifer Name	Lithologic Description	Thickness (ft)
Upper Surficial	Poorly graded sand to silt; clayey/silty sand; clean sand	0 - 25
Lower Surficial	Same as Upper	20 - 50
Upper Intermediate	Elastic silt with sand	30 - 80
Middle Intermediate	Limestone	
Lower Intermediate	Calcareous silt, limestone, carbonate sands; sandy silt	80 - 130
Upper Floridan	Calcareous clay; carbonate sands/silts	100 - 150

The top of the Floridan is typically at a depth of 120 to 140 feet below the ground surface. The major contaminants of concern in the groundwater are tetrachloroethene (PCE), trichloroethene (TCE), *cis*-1,2-dichloroethene (DCE), vinyl chloride (VC), toluene, and methylene chloride. Total volatile organics (TVOCs) range in concentration from 29 to 290,000 μg/L. Table 2 summarizes the various sites, aquifer depth, plume size and highest TVOC's at the start of groundwater remediation activities.

Table 2: Plume characteristics.

Site	Contaminant Plumes			
	Aquifer	Depth Interval (ft bls)	Plume size (acres)	Highest TVOCs (μg/L)
Landfill 1	Upper Surficial	0-25	4.5	160,000
	Lower Surficial	25-50	4.3	290,000
	Intermediate	90-100	2.3	20,000
	Floridan	100-150	2.7	74,000
Landfill 3	Surficial	0-20	9	46,000
	Lower Intermediate	125-145	5	2,100
Landfill 4	Surficial	0-20	5	7,300
	Upper Intermediate	45-60	8	40,000
	Lower Intermediate	80-130	0.5	2,500
Landfill 5	Surficial	30-50	19	18,000
	Upper Intermediate	58-68	10	740
	Lower Intermediate	100-110	2.7	410
	Floridan	125+	1.3	150
Landfill 6	Surficial	0-35	3.0	850
Launch test area	Surficial	0-25	0.6	3,200
	Upper Intermediate	30-60	0.6	2,300
Pershing/Waste conservation area	Upper Surficial	0-10	1	29
	Lower Surficial	20-25	11	12,000
	Upper Intermediate	40-50	25	5,800

Over 110 acres of contaminated area and 3.4 billion gallons of contaminated groundwater existed at the site at the inception of groundwater remediation activities.

2.2 Cleanup goals

At the UCPM South Campus site, a unique combination of approaches to attaining remediation goals was adopted. For solid materials, the overall objective of the corrective action was to meet regulatory requirements as a RCRA Interim Measures Corrective Action. This required removal of RCRA hazardous wastes and proper disposal offsite at RCRA permitted landfills. Other nonhazardous debris and soil was also disposed offsite or reused on site as backfill. Initially for groundwater, the overall objective was to achieve hydraulic containment and management of the groundwater plumes. The ultimate goal was to reduce concentration of the contaminants of concern to levels at or below State of Florida Groundwater Cleanup Target Levels (GCTLs). GCTLs are equivalent to State of Florida maximum contaminant levels (MCLs) which are typically lower than USEPA's MCLs. For example, the Florida MCLs for TCE and PCE are 3 μg/L and for vinyl chloride, the MCL is 1 μg/L (vs. 5 and 2 μg/L for Federal MCLs). After maintaining concentrations at of below GCTLs for one year, a request for no further action (NFA) or equivalent would be requested from the Florida Department of Environmental Protection.

2.3 Remediation

Remediation alternatives were evaluated based on effectiveness in meeting cleanup goals, implementability, costs and time to achieve the cleanup goal. In addition, remediation is being performed to facilitate phased property development in accordance with economically driven build-out plans and demand for the property.

2.3.1 Solid waste

Numerous innovative assessment and construction methods helped to sustain the dynamic and fast-track schedule. Development goals required excavation of abandoned landfills containing varying amounts of hazardous and nonhazardous materials. To support work plan development for landfill excavation, the project team conducted test-pitting with an onsite geologist and survey team. Using the resulting data, the team constructed digital, three-dimensional terrain models that served as the basis of design for landfill excavation and waste volume calculations. Because of the potential for unexploded ordinance, a full-time ordnance expert was assigned during all landfill excavation activities. Materials were stockpiled and screened using on site analytical methods to rapidly determine chemical characteristics and select proper disposal or reuse. Table 3 summarizes the quantities excavated, stockpiles characterized, and materials disposed or reused from each of the six landfills.

Over a four year period 417,275 cubic yards of materials were excavated from the six former landfills. In total over 25.5 acres up to a depth of 19 feet were

Table 3: Solid waste excavation and disposal summary.

Landfill	Landfill 1	Landfill 2	Landfill 3	Landfill 4	Landfill 5	Landfill 6
Excavation Dates						
Start	12/03/01	05/15/02	11/30/99	12/18/00	01/13/03	10/16/01
End	07/10/02	07/15/02	08/21/00	06/18/01	03/25/04	05/15/02
Duration (mo.)	7	2	7.5	7	14	7
Excavation Information						
Total Area Excavated (acres)	3.6	1.1	4.0	3.5	9.6	3.7
Total Volume Excavated (cu.yds.)	31,000	5,500	30,300	36,200	267,000	47,275
Excavation Depths (ft.bls)	2 - 8	2 - 9	2 - 7.5	2 - 10	2 - 19	3 - 11.5
Stockpiles Produced						
Debris	28	0	7	42	0	60
Soil	155	15	155	139	35	243
% Soil Used as Backfill	32%	33%	14%	40%	0%	70%
Wastes Disposed Off-Site						
F006 Sludge (tons)	5,183	4,534	1,382	2,076	3504	0
Nonhazardous Debris (tons)	1,851	0	548	7,827	0	9,578
Nonhazardous Soil (tons)	21,000	1,286	34,508	20,130	267,000	11,042
C&D Materials (cu.yds.)	780	234	1,260	4,550	4,500	2,576
Other Wastes (drums)	35 haz; 12 nonhaz	2 haz drums	6 haz; 8 nonhaz	6 haz; 9 nonhaz	0	1 nonhaz

removed. During the time, 137 stockpiles of debris and 742 stockpiles of soil were characterized. In all 16,679 tons of RCRA F006 sludge were disposed off site. In addition, 19,804 tons of nonhazardous debris and 354,966 tons of nonhazardous soil were disposed off site. Approximately 32 percent of the excavated soil was used on site as backfill.

2.3.2 Groundwater

Preliminary assessment of the nature and extent of the plumes was performed using direct-push technology with an onsite mobile gas chromatography/mass spectrometry (GC/MS) laboratory. This allowed the project team to accurately investigate large plume areas at varying subsurface depths in a relatively short time period. Once the preliminary plume assessment was complete, monitoring wells were placed in strategic locations to confirm results.

Using existing and previously collected data, a centralized groundwater treatment facility capable of processing groundwater from all contaminated areas at the site was designed and constructed. The treatment system consists of two 500-gallon-per-minute tray air strippers (design capacity of 1,000 gpm), a one million gallon receiving pond and an effluent disposal sprayfield. The effluent disposal sprayfield distributes treated groundwater over 80 acres, reducing treated water disposal costs and recharging the regional aquifers.

Groundwater recovery systems were designed and installed at each contaminated site and then connected to the centralized groundwater treatment facility (see Figure 1). Detailed modeling of groundwater flow, contaminant transport, and particle tracking was conducted to facilitate groundwater recovery system design. The groundwater recovery systems consist of over 4,450 linear feet of horizontal recovery wells and 41 vertical recovery wells. One of the horizontal wells is currently deemed the largest in the world in terms of depth (50 feet) using a single one-pass trenching machine. To date over 60 million gallons of contaminated groundwater have been recovered, treated to drinking water standards, and then returned to the aquifer system. A summary of the groundwater remediation systems is provided in Table 4.

Current extraction and treatment flow is approximately 280 gallons per minute. Supplemental technologies are in the process of being implemented to accelerate groundwater remediation in conjunction with pump and treat. These include open pit aeration, enhanced bioremediation, chemical oxidation and natural attenuation. For example, the following supplemental technologies are being implemented to aggressively remediate the following aquifers and plumes:

Landfill 3:

- Surficial aquifer: open pit in situ aeration followed by chemical oxidation (potassium permanganate)
- Lower Intermediate: five deep injection and two recovery wells for permanganate injection and recovery

Table 4: Groundwater remediation systems.

Site	Contaminated Plumes		Groundwater Recovery System Description			Design Rates	Comments
	Aquifer	Depth Interval (ft bls)	Recovery Wells Required	Well Total Depth (ft bls)	Startup Date	Gallons per Minute	
Landfill 1	Upper Surficial	0-25	27 air sparge wells	25	Feb-06	42	Source area investigation has shown free product areas totaling 0.09 acres in surficial aquifer that will require focused remediation.
	Lower Surficial	25-50	27 air sparge wells, 6 vertical wells	55/62			
	Intermediate	90-100	4 vertical wells	125			
	Floridan	100-150					Not installed at this time - MNA
Landfill 3	Surficial	0-20	4 horizontal wells	15	Oct-01	35	Each horizontal well is 325 feet long. System is in long-term monitoring.
	Lower Intermediate	125-145	1 vertical well	135			
Landfill 4	Surficial	0-20	2 horizontal wells	20	Jul-02	24	Each horizontal well is 350 feet long. System is in long-term monitoring.
	Upper Intermediate	45-60	4 vertical wells	50-70			
	Lower Intermediate	80-130	3 vertical wells	105-120			
Landfill 5	Surficial	30-50	3 vertical wells	45-55	Mar-06	108	90-Day Startup Scheduled for 2/06.
	Upper Intermediate	58-68	8 vertical wells	70-80			
	Lower Intermediate	100-110	3 vertical wells	105-110			
	Floridan	125+					Not installed at this time - MNA
Landfill 6	Surficial	0-35	2 horizontal wells	35	Apr-04	25	Each horizontal well is 425 feet long. System shutdown December 05.
Launch Test Area	Surficial	0-25	48 air sparge wells	25	Apr-05	0	Operate air sparge system until surficial aquifer is remediated before stating recovery wells in the lower zone.
	Upper Intermediate	30-60	9 vertical wells	60	May-05	9	
Pershing/ Waste Conserva-tion Area	Upper Surficial	0-10	4 horizontal wells	50	Oct-04	40	Each horizontal well is 400 feet long. Horizontal wells will operate for 1-2 years before startup of Floridan well.
	Lower Surficial	20-25					
	Upper Intermediate	40-50					

Landfill 4:

- Surficial: horizontal injection wells for permanganate distribution
- Upper Intermediate: 58 direct push wells for permanganate injection

As shown above, chemical oxidation is being combined with other effective techniques (pump and treat and open pit aeration) to accelerate remediation timeframes. Estimated cost to achieve no further action for groundwater at Landfills 3 and 4 using the above systems is approximately $3.6 million. Remediation is anticipated to be completed in 3 years (including one year of post-remediation monitoring) at target cleanup levels.

To monitor the efficiency of the recovery and treatment systems, an extensive data collection network was installed. Data are collected to evaluate remedial progress for regulatory reporting and to allows UCPM and its consultants to refine and prioritize site development strategies.

2.4 Redevelopment

To date over $70 million has been expended in total infrastructure costs at the site. Of this amount, approximately $35 million is related to remediation of contaminated materials and groundwater. Current estimates are that over $250 million will be expended for all infrastructure costs. The actual property investment value is estimated to be $2.5 billion. To date, the following development has occurred:

Expansion of the Orange County Convention Center
Pulte Homes Vistas at Lake Cay (Resort Rentals)
Intrawest Village of Imagine (Resort Rentals)
Universal Boulevard Expansion
Canadian Court Expansion
Rosen's Shingle Creek Resort
UCF Hospitality School

3 Conclusions

Approaches to redevelopment of contaminated properties vary depending upon many factors including type/extent of contamination, environmental/human health risk, regulatory compliance, remediation costs, compliance with Federal SEC reporting, the resources/desires of the owners/developers and the value and demand for the property. Because of the high costs and long time associated with typical groundwater remediation, innovative techniques to reduce groundwater remediation timeframes were employed including:

- Use of horizontal wells to effectively capture and recover contaminated groundwater.
- Use of chemical oxidation to rapidly cleanup the contaminated groundwater to target cleanup levels.
- Use of horizontal wells to distribute chemical oxidants.
- Creation and use of open pits to directly aerate the exposed groundwater *in situ* and remove contamination.

Reducing the remediation timeframe of groundwater is a common approach at many sites because groundwater remediation is typically the most expensive, time consuming and difficult media to ensure cleanup. In all cases, extensive evaluations and interaction with regulatory agencies were required to assure that the alternative approaches where protective of human health and the environment.

Other approaches and elements common that help result in successful development include:

- Integrated total site solutions allowing for cost saving approaches and synergism between individual areas of the site.
- Dynamic and real time investigations and to allow rapid characterization of contaminant levels and extent.
- On site and real time screening of excavated soils to allow rapid disposal and reuse decision processes.
- Flexible and dynamic remediation plans that could respond to the owners need to sell or develop property to maintain a balanced cash flow.
- Effective and rapid remediation technologies (e.g., chemical oxidation and open pit aeration) to facilitate property sale and development.
- Innovative and cost saving remediation techniques such as horizontal wells.
- Proactive coordination with regulatory agencies to build consensus and shorten review time of work plans, reports, etc.
- Large area in desirable location with great potential for development and revenue generation.
- Interim cleanup and development of selected areas while the remaining portions of the site were being investigated and remediated, which generated cash flow to allow site infrastructure construction and remediation measures implementation.

All of these approaches and attributes of the sites contributed to the ongoing successful remediation and redevelopment.

Brownfield sites turned green: realising sustainability in urban revival

K. J. Doick, G. Sellers, T. R. Hutchings & A. J. Moffat
Land Regeneration and Urban Greening Group, Forest Research, UK

Abstract

Greenspaces in the urban environment are widely recognised for their importance in the creation of healthy and sustainable communities. Greenspace establishment on reclaimed brownfield land is an important mechanism for reviving the urban environment. However, greenspace establishment impacts not only social but also environmental, ecological, economical and cultural dimensions. The identification of successes and failures within brownfield regeneration, through qualitative and quantitative evaluation, is paramount for demonstrating whether projects have attained their primary objectives, offer value for money and are sustainable. Currently, tools available for such evaluation fail to evaluate the true plethora of impacts of greenspace establishment and, hence, sustainability. This paper summarises over 110 identified impacts which could be used as evaluation criteria. The development and implementation of a toolbox based on the evaluation criteria is discussed, that would offer a means of determining monitoring needs on a site by site basis and a viable means of evaluating local or national impacts for a variety of urban greenspaces. The final product would provide a defensible mechanism by which redevelopments could be shown to have succeeded, according to a composite of individual functional aspirations and impacts. The lessons learnt through implementation of our toolbox would benefit future projects, enable Best Practice to be demonstrated and improved upon, and identify areas where benefits are not being fully realised.
Keywords: brownfields, urban regeneration, greenspaces, impacts, monitoring, evaluation, toolbox, social, environmental.

WIT Transactions on Ecology and the Environment, Vol 94, © 2006 WIT Press
www.witpress.com, ISSN 1743-3541 (on-line)
doi:10.2495/BF060131

1 Introduction

From as early as the second half of the 19th century, the provision of urban greenspaces in the form of municipal city parks has been seen as a means to alleviate the unhealthy attributes of city dwellers and factory workers Nicholson-Lord [1]. The open areas for physical exercise provided the additional benefit of being a place to socialise with friends and escape from the mental, as well as physical, confines of the work environment. Today, we have a much greater recognition of the benefits that urban greenspaces can offer and a realisation that they are not solely anthropomorphic (social, cultural, economic, planning, landscape) but also encompass environmental and ecological dimensions. The diverse range of benefits that urban greenspaces present is recognised in a UK government primary objective to "create safe and healthy local environments with well-designed public greenspace through a programme of brownfield land regeneration and community woodland establishment in the urban environment" ODPM [2]. Inherent in this objective is the consideration that brownfield redevelopment is *de facto* sustainable and, as such, has been presented as a UK government headline sustainability indicator DETR [3].

The drive for sustainable urban revival through brownfield redevelopment is not, however, solely in recognition of the benefits that urban greenspaces offer, but is a consequence also of economic pressures, demands for new housing and planning guidelines aimed at preventing urban sprawl. However, sustainability is not only about reducing consumption of raw materials or the reclamation of brownfield land as a sustainable alternative to developing virgin greenfield sites. Sustainability is an interplay between social, environmental, economic, cultural and ecological dimensions, over the long-term. Assessment of the sustainability of brownfield regeneration to urban greenspaces therefore requires sufficient evaluation of the impacts of the redevelopment and holistic interpretation of the interplay at work.

Examples of monitoring strategies currently available that can be used in the assessment of an urban greenspace establishment project, or in the design of an assessment package for such a project include the *Redevelopment Assessment Framework* (RAF) [4], *A Guide to Monitoring and Evaluation* [5], and *Prove it*! [6]. These packages only partially fulfil the monitoring specifications required to appraise the sustainability of an urban greenspace regeneration project. This is because each addresses only a limited number of evaluation criteria in their assessments. For example, RAF considers stakeholder interests but does not evaluate the regional economic implications of the redevelopment, whilst methodologies such as the UK government's *Green Book* (which is an approach to the evaluation of public sector expenditure on activities aimed at providing benefits to society [7]) do not consider wider environmental aspects, such as changes in ecological status.

This paper summaries results of a literature review that identified over 110 impacts of establishing greenspaces in the urban environment, which could be used to formulate evaluation criteria for the assessment of project sustainability. The development and implementation of a toolbox based on the evaluation criteria, together with stakeholder analysis, is discussed, that would allow monitoring needs to be determined on a site-by-site basis. Evaluation of the monitoring data would also offer a viable means of understanding local, regional and national impacts of brownfield regeneration-greenspace establishment projects. Employable at a range of greenspace types, from recreational playing fields to community woodlands, the toolbox would provide a defensible mechanism by which redevelopments could be shown to have succeeded, according to a composite of individual functional aspirations and impacts. The lessons learnt through implementation of our toolbox would provide the evidence base to support and benefit future projects, enable Best Practice to be targeted, demonstrated and improved, and identify areas where benefits are not being fully realised.

2 Impacts of establishing urban greenspaces

2.1 Defining the impacts of establishing urban greenspaces

Table 1 summaries an extensive literature review into the impacts associated with greenspace establishment in the urban environment. The review identified over 110 impacts related to social, environmental, ecological, engineering, economical and cultural aspects of urban greenspace regeneration.

Table 1: Impacts of establishing urban greenspace on brownfield land.

Generic Impact	Details	Summary of Evidence
Social	General public visits Sports and recreational facilities Human health, health and safety Access, sense of ownership Social inclusion (array of social classes using site, evidence of social exclusions) Educational resource Fear (real, perceived)	Numbers of: Formal visits (social activities/ events) Informal visits (passive enjoyment; walking the dog, socialising) Increase in personal / public health and well-being (active enjoyment, mental well-being and release) Extent of open public access Too many/too few people around
Urban impacts	Urban heat island effects (ambient temperature, cooling due to evapotranspiration and shading); Meteorology Noise abatement Aesthetic appeal of city/urban environment Visual screening of urban structures	Impacts on people's thermal comfort Atmospheric pollution formation cycles Economics in relation to heating and air conditioning bills [8–10]
Air quality (air pollution)	Particulates (PM_{10} and $PM_{2.5}$) Heavy metals and potentially toxic elements NO_x and SO_x, Ozone Biogenic and anthropogenic hydrocarbons, persistent organic pollutants, volatilization of bitumen Air temperature	Trapping and uptake of atmospheric pollutants by vegetation VOCs emitted as well as trapped by vegetation [11, 12]

Table 1 Continued.

Generic Impact	Details	Summary of Evidence
Soil and ground contamination issues	Compartmentation of contaminants in soil and vegetation Potential of food chain transfer Exposure to human visitors Effects of re-vegetating on contaminant stability Erosion extent and impact	Contaminant uptake by plants leading to food chain transfer Sub-standard vegetation establishment due to contaminants and low quality status of soil Penetrate of landfill caps by tree roots [13, 14]
Water supply, management, quality and chemistry	Rainfall interception and infiltration Surface water quality (pH, BOD, nitrate, suspended solids, turbidity) Invertebrate biology Water reserves (site and region) Flood risk and management	Infiltration rates influence catchment yields, on-site and off-site flooding potential Soil runoff affects water quality and loading [13, 15]
Soil quantity/quality	Soil chemistry (organic matter, soil carbon, NPK, pH, CEC) Biology (invertebrate diversity and numbers) Physical (density, temperature)	Low quality soil resource, low organic matter content, poor nutrient status, minimal faunal populations – temporal changes post regeneration Vegetation reduces soil erosion.
Biodiversity and conservation value	Tree (health, vigour, damage, stand management, weed control) Flora (habitats, species composition, diversity, performance) Fauna (invertebrate through to reptile and mammal)	Low quality soil resource affects tree and vegetation establishment Temporal changes in soil flora and fauna, aquatic and total biodiversity [15–17]
Habitat creation	New and native woodland creation, wetlands, wildflower meadows and parklands Natural regeneration	Site sustainability affected by site capacity to self regenerate National sustainability measured by coverage of native habitats
Engineering aspects	Soil compaction, water-logging Landfill tip stability Engineering structures functionality (e.g. berms, drains and culverts) Road/path condition (fitness for purpose) Evidence of leachate breakout	Compaction limits rainfall infiltration and vegetation establishment Durability (wear and tear as well as vandalism) of sub surface and on site engineering structures indicative of sustainability and long term maintenance requirement of site
Site condition	Aesthetic appeal Landscape value Littering and vandalism Nuisance (motor bikes)	Aesthetic appeal influences site usage through to desirability of neighborhood and local house prices Vandalism indicates financial burden of site and sense of ownership by local community [18, 19]
Economic	Employment creation (recreation, sport, tourism, site maintenance, management, influx of new business) Increased local prosperity (land value, existence value, jobs) Carbon sequestration Non-marketable benefits (pollution mitigation, personal health, amenity)	Revenue income from users Value of peripheral land and property Trees and plants as a marketable resource [20–22]
Cultural	Cultural identity Interaction, use by ethnic groups Landscape value Preservation of environmental/ natural and cultural heritage, historical heritage and archaeology	Historical and industrial uses of a site relate to local community, ranging from proud of heritage to site as an eyesore. Sense of ownership, belonging and pride [16, 18–20]

2.2 Capturing the impacts of establishing urban greenspaces

The impacts associated with urban greenspace establishment demonstrated in Table 1 are wide ranging. Some impacts are subtle (e.g. changes in the numbers of dog walkers visiting a site), whilst others are more dramatic (e.g. changes in aesthetic appeal of the site). Some occur immediately upon regeneration (e.g. availability of recreational space for the local communities), others may take years to decades to be fully realised (e.g. attitudes of local communities to the environment, or sense of ownership). Consequently, resource requirements to qualify or quantify these impacts vary considerably. For example, data collection can proceed in a variety of ways from site surveys by local interest groups (e.g. wildlife groups) through to interpretation of national databases. How then do we assure data quality and comparability over the evaluation period?

Quality assurance is imperative not only for ensuring that data is comparable between sites, but also to prove that decisions and conclusions for each individual site are fully accountable and auditable. Therefore a formal and structured quality management system (QMS) must be adopted. A QMS such as the UK's *Joint Code of Practice for Research* [23] requires an overarching deployment plan that unambiguously defines the project concept, a review of the literature and the main aims and objectives of the project. Standard operating procedures (SOPs) are subsequently produced, which enable any competent person to undertake monitoring (data collection) at any site in a manner that will generate data that is comparable (by virtue of the standardised data collection methodology) to other sites in the same regeneration programme as well as to unrelated and control sites. Furthermore, a data management system (DMS) is required, in support of the QMS, which offers a central secure location for data to be analysed following standard procedures for collation, manipulation, transformation and statistical analysis.

3 Formulating a monitoring and evaluation strategy

3.1 Traditional ad hoc approaches

Whether for one site or a regeneration programme consisting of multiple sites, assessment of the sustainability of the project can only occur through a truly holistic appraisal of the regeneration process and the impacts that ensue. There are several approaches that are potentially useful in deciding what criteria should be included in the monitoring and evaluation strategy, given that the range of impacts is so diverse. A traditional approach is to list the primary goals, outputs and aspirations for a site. Subsequently, monitoring is targeted directly and exclusively at those items on the list to demonstrate the successes and short-comings of the project. Such an approach does little to capture the wider impacts of the project, or evaluate site sustainability post-regeneration. For example, the primary objectives of a regeneration project may be to provide public benefit through a variety of recreational facilities and improved local connectivity through the provision of new foot and cycle paths. If these objectives were the sole items for monitoring and evaluation, wider impacts such as increased site (local) biodiversity would not be captured.

An alternative approach to monitoring and evaluation is via interpretation of the data needs of the funding body or principal stakeholder(s); in other words, via stakeholder analysis. For example, regional development agencies in the UK have a specific role of promoting economic regeneration. Subsequently, a regeneration project would include primary objectives such as 'create employment opportunities', 'complement other regeneration activity' or 'make significant contributions to the delivery of the regional economic strategy'. A monitoring strategy devised to exclusively demonstrate the fulfilment of these objectives may offer insight into the numbers of new businesses set up in the region, the influx of skilled workers or changes in house prices in the vicinity of the project, but is unlikely to demonstrate environmental consequences such as changes in bird of prey or mammal numbers, atmospheric pollution abatement or the attitudes of local people to the site (e.g. using quality of life indices [24]). The assumption that regeneration is inherently sustainable development may prove to be flawed if monitoring results demonstrate negative externalities, low-value project returns or lower than expected returns for the capital investment, or if factors extraneous to the regeneration programme (e.g. a slump in house prices) mask any regional economic growth promoted by the greenspace development.

3.2 An integrated and holistic approach to monitoring and evaluation

Monitoring and evaluation strategies may encompass several of the primary objectives or goals for a regeneration project, the functions or roles expected to be fulfilled by the greenspace, site aspirations or project outputs. To be useful, monitoring the sustainability of a regeneration project needs to be via a strategy bespoke to each site, but how does one have a bespoke design yet still maintain data comparability across sites in a rigorous QA system? A review of urban greenspace establishment programmes and of the literature has shown that there is significant and relevant overlap between the objectives of different regeneration projects; equally, stakeholder interests are very similar between the projects. Furthermore, greenspaces fulfil often similar functions for local communities, and similar impacts of regeneration can be identified even between dissimilar locations and sites. It is the recognition of these overlaps, and the ability to rank and sort potential criteria for monitoring that enables a strategy to be designed that meets approval by the funding bodies and is capable of evaluating site sustainability.

Figure 1 presents a flow diagram demonstrating the process that should be followed in order to select a balanced list of evaluation criteria for a regeneration project; monitoring indicators are subsequently selected from the toolbox to satisfy the identified evaluation criteria. The first step involves identification of stakeholders, project aims and objectives, functions and impacts. Stakeholders are identified from a checklist within the toolbox, which includes *inter alia* statutory and non-statutory bodies, local interest groups and local communities, public bodies and funding bodies. The conversion of stakeholder views, aims, aspirations, concerns and interests into relevant evaluation criteria is undertaken via a sequence of workshops, wherein stakeholders state, discuss and then collectively agree their principal criteria for evaluation. An example of how this process may be undertaken is given by Pediaditi *et al.* [4].

Project aims and objectives are identified from the project outline, proposal or funding application. Functional aspirations are revealed through public consultations documentation (surveys and minutes of meetings), pre-development ecological site surveys, desktop studies and site walkover reports. Finally, the impacts of a greenspace establishment programme are assessed via the toolbox. Site factors such as history and location play an important role in the selection of probable impacts of the project that should be considered. However, a range of impacts from the toolbox, relating to all the dimensions (social, environmental, ecological, economical and cultural, as well as local, national and regional impacts), should be selected to ensure an integrated approach to monitoring and the potential to identify unexpected impacts. In this way, the toolbox reflects the whole programme approach to monitoring advocated by European Directive 2001/42/EC '*on the assessment of the effects of certain plans and programmes on the environment*' [25] and the integrated approach to *Sustainability Appraisal* endorsed by ODPM [26].

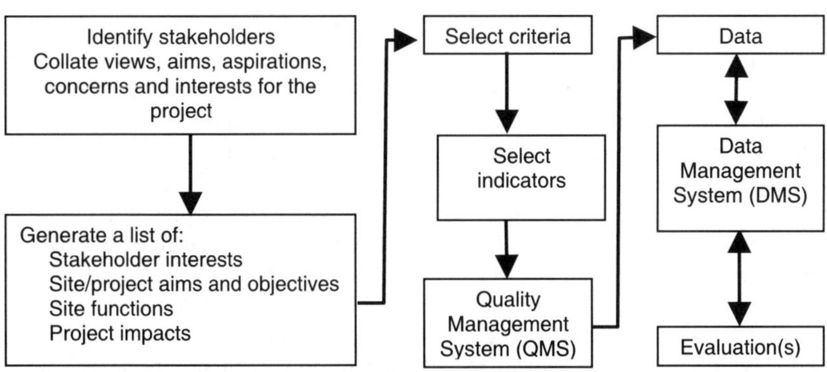

Figure 1: Selection of evaluation criteria and monitoring indicators based on stakeholder interests, site functions and impacts of urban greenspace.

For the toolbox, a QMS is proposed that is compliant with the *UK's Joint Code of Practice* [23]. Subsequently, the QMS would require an over-arching deployment plan that details the concept of the toolbox and its role and purpose. The QMS will also demonstrate how the over-arching deployment plan applies at local, regional and national scales, and how deployment plans should be drawn up for each new site. These plans will derive the evaluation criteria (i.e. impacts, aims, functions and stakeholder requirements) specific to the programme to be monitored and which indicators will be employed for each criterion. SOPs for data collection and handling will then be taken directly from the toolbox for use at each site to ensure that methodologies are consistently applied.

A concern when devising any monitoring and evaluation strategy is resource availability; inevitably financial constraints limit the number of monitoring indicators that can be assigned. It is therefore pertinent to stress that project aims and objectives are frequently the same or very similar to, for example, site functions or stakeholder data requirements – it is just that they are, invariably, listed separately. Similarly, many impacts of a project will also be intentional project functions, outputs or outcomes. Through systematic identification and, subsequently, detailing the aims and functions, etc., of a project, the maximum number of evaluation criteria can be identified and the optimum number for adoption can be defensibly agreed. Importantly, it is this initial, systematic approach that proves the maximum list and, therefore, ensures a holistic approach; in other words, important criteria are not omitted through oversight. The toolbox ensures that at least one indicator is selected per evaluation criterion and only fitness-for-purpose indicators are employed. Indicator suitability will be tested against parameters of *inter alia* specificity, responsiveness and natural variability, measurement error and longevity (Hunsaker [27]).

4 Implications of the monitoring and evaluation toolbox

The identification of successes and shortfalls in greenspace establishment projects, through qualitative and quantitative evaluation, is paramount for demonstrating whether projects have attained their primary objectives and delivered value for money. The need for such evaluation is recognised by Governments, yet the few disjointed tools that are currently available fail to evaluate the true plethora of impacts associated with either regeneration projects *per se* or the establishment of greenspace. The identification of the multitude of impacts of a regeneration project enables conclusions to be drawn from the evaluation process that provide a well balanced picture of the sustainability of the project. Undeniably required, the proposed toolbox discussed herein will provide a defensible mechanism by which redevelopment programmes can be shown to have succeeded. Bespoke monitoring and evaluation strategies, on a site or programme basis, will enable success to be measured according to a composite of individual functional aspirations, directly relevant to that site. Concurrently, the meticulous design of a bespoke monitoring strategy ensures that stakeholder demands for project feedback are satisfied and resources are targeted at areas most in need of monitoring. Additional benefits of a bespoke strategy include the ability to make accountable management decisions and map temporal changes in outputs, impacts or functions that are primary objectives for a site. Finally, by ensuring data are fully comparable between sites and through widespread application of the monitoring and evaluation strategies designed using our toolbox, lessons can be learnt to benefit future projects, enable Best

Practice to be demonstrated and improved, and changes can be implemented where social, environmental or economical benefits are not being fully realised.

References

[1] Nicholson-Lord, D., *The Greening of the Cities*. London: Routledge and Kegan Paul, 1987.

[2] Office of the Deputy Prime Minister (ODPM), *Sustainable communities: building for the future*, ODPM London, 2003.

[3] Department of Environment, Transport and the Regions (DETR), *A better quality of life: A strategy for sustainable development in the UK, Cm 4345*, The Stationery Office, London, 1999.

[4] Pediaditi, K., Wehrmeyer, W. & Chenoweth, J., Monitoring sustainability of brownfield redevelopment projects: the Redevelopment Assessment Framework. *Land Contamination and Reclamation*, **13(2)**: pp. 173-183, 2005.

[5] ECOTEC, Open Space and The Countryside Agency, *A guide to monitoring and evaluation*, The Countryside Agency: London, p. 24, 2004.

[6] New Economics Foundation, *Prove it! Measuring the effect of neighbourhood renewal on local people*, Groundwork, London: RAP, 2000.

[7] HM Treasury, *Green Book: appraisal and evaluation in central government*, The Stationery Office, London, 2003.

[8] Brack, C.L., Pollution mitigation and carbon sequestration by an urban forest. *Environmental Pollution*, **116(Supplement 1)**: pp. S195-S200, 2002.

[9] Akbari, H., Shade trees reduce building energy use and CO_2 emissions from power plants. *Environmental Pollution*, **116(Supplement 1)**: pp. S119-S126, 2002.

[10] Fang, C.-F. & Ling, D.-L., Guidance for noise reduction provided by tree belts. *Landscape and Urban Planning*, **71(1)**: pp. 29-34, 2005.

[11] Taha, H., Modeling impacts of increased urban vegetation on ozone air quality in the South Coast Air Basin. *Atmospheric Environment*, **30(20)**: pp. 3423-3430, 1996.

[12] Beckett, K.P., Freer-Smith, P.H. & Taylor, G., Urban woodlands: their role in reducing the effects of particulate pollution. *Environmental Pollution*, **99(3)**: pp. 347, 1998.

[13] Hartanto, H., Prabhu, R., Widayat, A.S.E. & Asdak, C., Factors affecting runoff and soil erosion: plot-level soil loss monitoring for assessing sustainability of forest management. *Forest Ecology and Management*, **180(1-3)**: pp. 361, 2003.

[14] Nowak, D.J., Crane, D.E., Walton, J.T., Twardus, D.B. & Dwyer, J.F., Understanding and quantifying urban forest structure, functions, and value. *Proc. of the 5th Canadian urban forest conference*, eds. W.A. Kenney, J. McKay, and P. van Wassaneaers, Urban Forest Council: Markham, Ontario, Canada, 2002.

[15] Freer-Smith, P.H. & Read, D.B., The relationship between crown condition and soil solution chemistry in oak and Sitka spruce in England and Wales. *Forest Ecology and Management*, **79(3)**: pp. 185, 1995.

[16] Dunnett, N., Swanwick, C. & Woolley, H., *Improving Urban Parks, Play Areas and Open Spaces*, ed. Department for Transport, Local Government and the Regions (DTLR), The Stationery Office, London, pp. 217, 2002.

[17] Biodiversity in Urban Habitat Patches: URGENT Annual Meeting 2000 Proceedings: Ecology Science Project Presentations; Angold, P., Hill, M., Rushton, S. & Pullin, A., Online. http://urgent.nerc.ac.uk/Meetings/2000/2000Proc/ecology/angold.htm

[18] Nijkamp, P., Rodenburg, C.A. & Wagtendonk, A.J., Success factors for sustainable urban brownfield development: a comparative case study approach to polluted sites. Special Section: Economics of Urban Sustainability. *Ecological Economics*, **40**: pp. S235-S252, 2002.

[19] Kuo, F.E. & Sullivan, W.C., Environment and crime in the inner city: does vegetation reduce crime? *Environment and Behavior*, **33(3)**: pp. 343-367, 1999.

[20] De Sousa, C.A., Turning brownfields into green space in the City of Toronto. *Landscape and Urban Planning*, **62(4)**: pp. 181, 2003.

[21] Gatrell, J.D. & Jensen, R.R., Growth through greening: developing and assessing alternative economic development programmes. *Applied Geography*, **22(4)**: pp. 331, 2002.

[22] Thériault, M.K.Y. & Des Rosiers, F., The impact of mature trees on house values and on residential location choices in Quebec City. *Proc. of the. 1st biennial meeting of the International Environmental Modelling and Software Society*, eds. Rizzoli, A.E. & Jakeman A.J., University of Lugano, Switzerland, pp. 478-483, 2002.

[23] Department for Environment, Food and Rural Affairs (DEFRA), *Joint Code of Practice for Research*, 2003, Online. http://www.defra.gov.uk /science/documents/QACoP_V8.pdf

[24] Department for Environment, Food and Rural Affairs (DEFRA), *Securing the Future - UK Government sustainable development strategy*, The Stationery Office, London. pp. 186, 2005.

[25] European Commission, *On the assessment of the effects of certain plans and programmes on the environment*, The Strategic Environmental Assessment [SEA] Directive, 2001/42/EC, 2001.

[26] Office of the Deputy Prime Minister (ODPM), *Sustainability appraisal of regional spatial strategies and local development frameworks,* Consultation paper, 2004, Online. www.odpm.gov.uk/stellent/groups/odpm_planning/documents/page/odpm_plan_030923.pdf

[27] Hunsaker, C.T., New concepts in environmental monitoring: the question of indicators. *The Science of the Total Environment Supplement*, **Part 1**: pp. 77-95, 1993.

Toward the integration of brownfield rehabilitation and planning methodologies: case study of Keihin Industrial Area, Tokyo, Japan

A. Murayama[1], K. Banno[2], S. Ishii[1], T. Kurose[3] & T. Sato[2]
[1]*Center for Sustainable Urban Regeneration, School of Engineering, The University of Tokyo, Japan*
[2]*Engineering & Risk Services Corporation, Japan*
[3]*Department of Urban Engineering, School of Engineering, The University of Tokyo, Japan*

Abstract

Keihin Industrial Area (KIA) is the largest industrial area in Japan that has supported the country's rapid economic growth in the twentieth century. Now KIA calls for the integrated methodology of brownfield regeneration and planning as it gradually changes from the heavy industrial base to a new urban area that potentially contains research and development institutions, light industry, business and commercial facilities, housing and other uses. The paper, after introducing the current framework for brownfield rehabilitation in Japan and the current planning of KIA, presents key factors to be considered in establishing an integrated methodology for KIA regeneration.
Keywords: Tokyo, brownfield rehabilitation, land use planning, methodology.

1 Introduction

Tokyo is the largest urban region in the world containing more than 34 million inhabitants in the four prefectures of Tokyo, Kanagawa, Chiba and Saitama [1] (Figure 1). Industrial areas of various sizes are dispersed throughout the region, but the largest concentration is in Tokyo Bay Waterfront Area that holds Keihin Industrial Area (4,400 ha) and Keiyo Industrial Area (4,700 ha). These industrial areas were the engines of Japan's economic growth in the twentieth century, but

WIT Transactions on Ecology and the Environment, Vol 94, © 2006 WIT Press
www.witpress.com, ISSN 1743-3541 (on-line)
doi:10.2495/BF060141

they are now experiencing a gradual change as they entered the globalizing twenty-first century. Figure 2 shows the change of production by major types of industry in Keihin Industrial Area. The production of the industrial area hit its peak in 1985 and is dropping afterwards. Similar trend can be observed in Keiyo Industrial Area. Under these circumstances, many parts of Tokyo Bay Waterfront Area have been redeveloped to mixed-use (i.e. residential and commercial) urban areas such as Makuhari, Urayasu, Odaiba, Shiodome, Tennouzu and Minatomirai, or to residential areas such as Toyosu, Shibaura and Konan (Figure1). However, the vast industrial areas still remain.

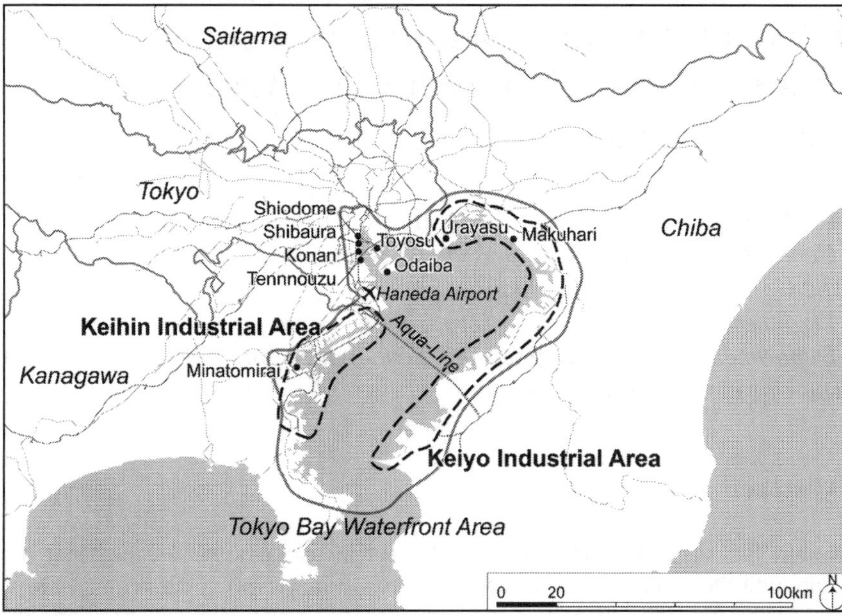

Figure 1: Map of Tokyo region with Keihin and Keiyo Industrial Areas.

This study focuses on City of Yokohama part (2,200 ha) of Keihin Industrial Area (KIA), one of the areas that have not been redeveloped to either mixed-use or residential areas. After the opening of the Port of Yokohama in 1859, KIA was developed through the private reclamation projects in the 1920s by a businessman Soichiro Asano. After World War I, major cement, steel, electronics, oil, wharf and warehouse companies located their plants in KIA. These plants, while driving Japan's economy, are thought to have left behind various hazardous substances in the soil. KIA kept on expanding even after the World War II until the completion of Daikoku Harbor in 1990.

Recently, there has been a drastic movement among companies in KIA such as the mergers of oil and steel companies or relocation of plants to foreign countries, which results in the generation of potential sites for redevelopment. On the other hand, research and development institutions have been located in KIA such as RIKEN Yokohama Institute and Yokohama City University

International Graduate School of Arts and Sciences in Suehirocho, introducing new land uses to the area.

Thus KIA calls for the integrated methodology of brownfield regeneration and planning as it gradually evolves from the heavy industrial base to a new urban area that potentially accommodates research and development institutions, light industry, business and commercial facilities, housing and other uses.

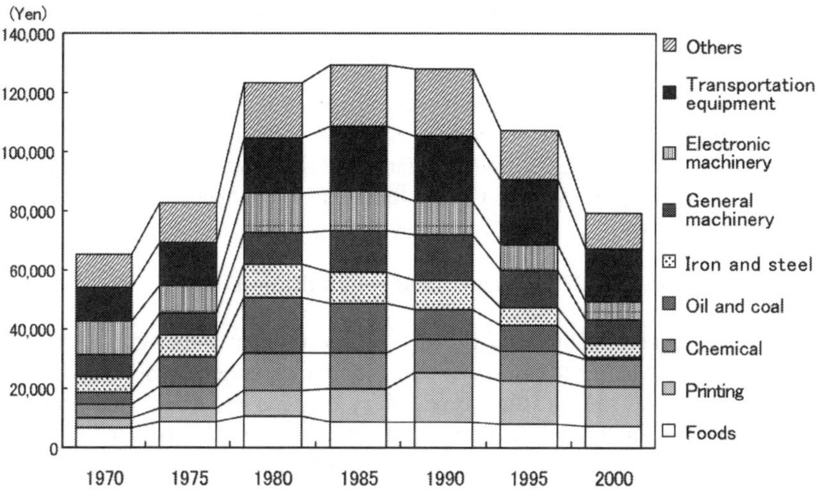

Figure 2: Production by major types of industry in Keihin Industrial Area [2].

2 Current framework for brownfield rehabilitation

"Brownfield" is not officially defined in Japan. To avoid misunderstanding of this technical term, it is defined for this paper as "contaminated land potentially to be developed".

2.1 Soil Contamination Countermeasures Law and local regulations

From the latter half of the 1980s and to the 1990s, buried industrial wastes and the surrounding contaminated soil were often unearthed at closed factory sites when they were redeveloped to residential apartments or commercial complexes. In response to this situation, Government of Japan set up a series of legislative measures such as the National Soil Environmental Standards in 1991, the Guidelines for Site Assessment and Response Actions in 1996 and the Soil Contamination Countermeasures Law (SCCL) in 2003.

Under the SCCL, there are two cases where landowners are obligated to carry out an environmental site assessment and remediation. Firstly, the owner of a hazardous wastewater treatment facility, normally registered to prefectural environmental administrator's office, has to conduct a site assessment when closing the facility. Secondly, the owner of a contaminated land property needs to take responsible actions when human health is threatened by polluted groundwater spreading from the site to the neighbouring environment. There is an option only for an industrial manufacturing site owner, however, to postpone conducting a site assessment in case where the land use remains the same for the future development.

In addition to the national regulation mentioned above, there are local regulations implemented by municipal statutory authorities. The purpose of local regulation is not to deregulate the SCCL but to help provide more opportunities or create more situations to require site assessments or to force landowners to disclose appropriate site information from the risk communication perspective. City of Yokohama covering a part of KIA has its own regulation. This requires: 1) analyses of dioxins in addition to the hazardous substances designated under the SCCL and 2) manufacturing companies to conduct site assessments and report to the city when they plan to conduct excavation works or close their factories.

2.2 Effects of the SCCL on brownfield rehabilitation

The survey conducted by Ministry of the Environment, Japan reports that the number of sites where site assessment was conducted under the SCCL was 93 [3] (Table 1), whereas the statistics provided by the Geo-Environmental Protection Center (GEPC) shows that the number of site assessments conducted by private contractors was as many as 4,281 [4], both of which were compiled for Fiscal Year 2003. The number reported by GEPC includes the sites unreported to any governmental agencies and the SCCL sites. This comparison suggests that the SCCL data represents only a small portion of contaminated or potentially contaminated sites, implying that the amendment of the law may be required in the future.

Table 1: Assessments conducted under the SCCL in FY 2003.

SCCL assessments	Number of cases
Facility closures that trigger SCCL assessments	590
Completed SCCL assessments	63
(The number of contaminated sites found)	(22)
Ongoing SCCL assessments	30
Pended SCCL assessments	491

The SCCL basically requires a “contain and control” approach for soil contamination countermeasures. For example, when a land is contaminated with heavy metal compounds, the landowner may be required first to construct asphalt pavement or soil cap layer on top of the contaminated soil to prevent harmful dusts from being inhaled by humans. If there is a risk of metal leachate migrating beyond the property boundary, groundwater flow must be controlled by the installation of sheet piles or the equivalent, and the contamination levels must be monitored to guarantee the effectiveness of “contain and control”. As for volatile organic compounds such as benzene or trichloroethylene, the SCCL requires groundwater controls. Vapour coming out from extended plumes has not been addressed under the SCCL so far since its risk to human health has not been well examined.

The SCCL has also introduced a grant that provides funding for a site assessment and restoration in case where a landowner is not a polluter and no identified polluter is affirmed to exist. This incentive has a great potential to encourage brownfields rehabilitation. However, it has never been utilized since the SCCL was enacted.

2.3 Current situation on site cleanup

Although the SCCL basically requires “contain and control” approach for countermeasures, more than 80 percent of the countermeasures that have been adopted in real site restoration ended up in a “dig and haul”, according to the survey conducted by Ministry of the Environment [3] (Table 2).

Table 2: Countermeasure methods (multiple answers possible).

Countermeasure methods	Number of sites
Groundwater monitoring	300 (6)
Soil contamination removal	1127 (17)
In-situ containment	60 (1)
Liner sheet installation	2 (0)
In-situ solidification/stabilization	56 (0)
On-site stabilization	45 (0)
Concrete vault containment	35 (0)
Soil capping	66 (1)
Pavement	122 (2)
Off-limits advisory	58 (2)
Others	301 (0)
The total number of respondents	1307 (19)

Note: The numbers in the parentheses represent the SCCL sites.
This table is assembled with the survey results from1991 to 2003.

The survey report doesn’t explain why soil contamination removal became such a widely adopted countermeasure method, but the following factors seem to support this cleanup trend: 1) The contaminated soil recycling business have been established, using the well-established domestic land and sea transportation networks, among the soil treatment facilities which utilize separation, evaporation or combustion methods, and 2) most of the former industrialized area in and around the urban areas have been converted into residential area, where prospective purchasers tend to be aware of stigma and to expect so called “complete cleanups”.

A "dig and haul" approach is not necessarily the best solution to make brownfield redevelopment cost-effective and environmentally and socially beneficial to communities. More collaborative works should be carried out between planners and environmental specialists to share the comprehensive scopes and to design appropriate contaminated site rehabilitation schemes.

3 Planning and existing conditions

3.1 Past and current regional and local planning

Many planning studies for KIA regeneration have been conducted since the 1970s. In the 1970s, several studies were conducted on Tokyo Bay Aqua-Line, the trans-bay expressway that connects Keihin and Keiyo Industrial. No studies on redevelopment have been conducted in this era of rapid growth. In the 1980s, National Land Agency (now integrated to Ministry of Land, Infrastructure and Transport), Government of Japan, have conducted numerous studies on redevelopment in Tokyo Bay Waterfront Area. The government at that time was promoting urban redevelopment to boost the country's economy and various areas in Tokyo Bay Waterfront Area were among the targets. Many of the new mixed-use urban areas and residential areas were planned in the 1980s, slowly developed in the recession of the 1990s and completed around 2000.

Corresponding to such regional planning studies on urban redevelopment by the national government, City of Kawasaki and City of Yokohama partly supported by the national and the prefectural governments initiated several local planning studies to set a new vision for KIA and to implement necessary projects. The major characteristic of the studies by the cities is that they focused on industry-related redevelopment and infrastructure projects instead of urban redevelopment projects. The cities have been promoting industry-related growth to aim for more tax revenues.

The latest plans for KIA are Keihin Waterfront Area Regeneration Framework [5] and Keihin Waterfront Area Regeneration Master Plan [6]. The framework stressed on (1) industrial revitalization by promoting new industries, developing Foreign Access Zone (FAZ) facilities for foreign trades and attracting businesses to underutilized sites, and on (2) infrastructure projects including the efficient use of the existing freight railway tracks, the improvement of the metropolitan expressway and the development of green spaces for disaster prevention. The master plan set out more detailed policies for the following six zones within Yokohama City limit of KIA (Figure 3).

Zone 1: Inland urban regeneration in cooperation with waterfront regeneration
Zone 2: Promotion of land use transition to create a mixed-use urban area
Zone 3: Advanced manufacturing center with global competitiveness
Zone 4: Advanced research and development center related to manufacturing
Zone 5: General distribution center
Zone 6: Advanced and efficient manufacturing center.

It also set out issue specific policies for railway and roads, for creating attractive space including historic conservation, industrial tourism and landscaping, and for improving performance in case of disaster. Furthermore, the following three districts are designated as priority districts.

Shinurashima and Moriyacho: Promotion of land use transition to create a mixed-use urban area by improving railway and road infrastructure, creating job opportunities and landscaping rivers and canals.

Daikokucho: The deregulation of industry-related restrictions and the improvement of arterial roads to further improve existing industrial functions and to enhance global competitiveness.

Suehirocho: Development of a new research and development center by attracting advanced small-medium size businesses and research institutions, effectively using existing freight and passenger railway tracks, improving road network and landscaping waterfront.

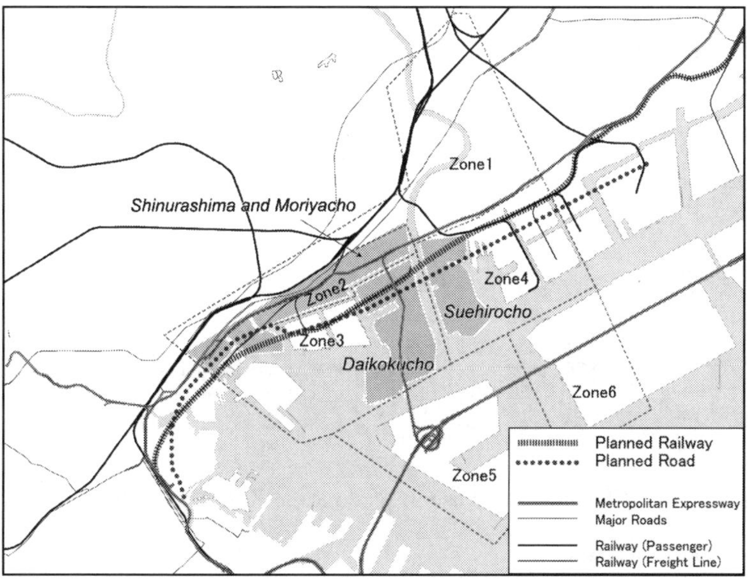

Figure 3: Six zones, three priority districts and improvement of railway and road infrastructures [6].

Thus, broad framework and issue specific policies for Yokohama City part of KIA, more detailed policies for the six zones within and development policies for the three priority districts are established through the planning efforts. However, explicit objectives such as employment and population targets, environmental standards, level of transportation services and visual image of urban design for the whole area are not yet clear to promote redevelopment activities. Objectives and images are considered on an individual project basis. To set forward the regeneration of KIA, the authors think it necessary to develop a comprehensive plan which includes the elements of industrial development, land use, transportation, building types and urban design of public space.

3.2 Existing conditions

3.2.1 Land use

Figure 4 shows that six zones of KIA (City of Yokohama part only) lost more than 175 ha of industrial uses from 1985 to 1995. In zones 1 and 2 bounded on inland urban area, industrial uses are changed to residential or commercial uses. In zones 3 and 5, the increase of freight uses can be observed. The increase of vacant sites is also significant adding up to approximately 105 ha in 2002. Figure 5 shows the distribution of twelve underutilized sites as of 2002 with the proposed land use transition. City of Yokohama proposed these directions for land use transition based on the recognition that nanotechnology, biotechnology and information technology are the new industries that can support KIA, but this does not impose any land use or redevelopment restrictions.

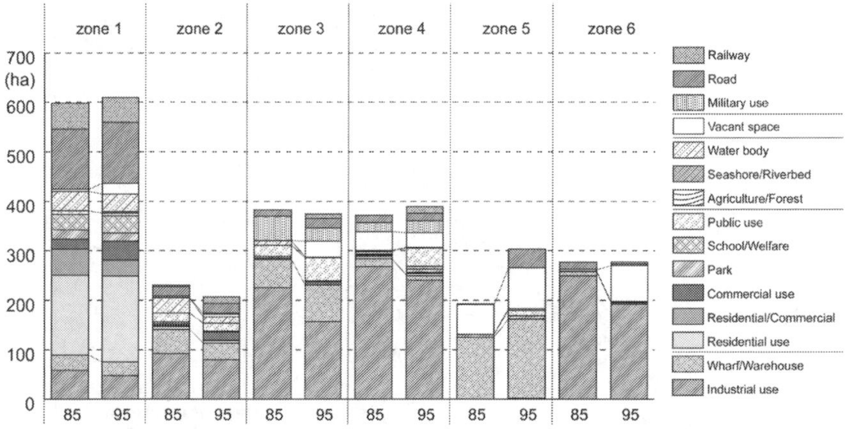

Figure 4: Change of land use in six zones of KIA, City of Yokohama [7].

3.2.2 Transportation

Public transport in KIA is limited. Passenger railway constructed mainly for factory workers connects some parts of zone 4 with JR (Japan Railway) Tsurumi and Hamakawasaki stations, but the service is limited compared to average passenger railway services in Tokyo region. To improve access by railway, utilization of the existing freight tracks for passenger services is under consideration, but the discussion has been stagnant since the area's land use plan is still not clear enough. On the other hand, arterial roads for cars and trucks have already been constructed and will be extended in the future including the connection to Yokohama Ring Road, Route 357, Tsurumi Waterfront Arterial Road and several connections to the inland urban area. These road extensions are needed to mitigate existing traffic congestions and accommodate more traffic in the future (Figure 3).

3.2.3 Canals, green spaces and historic structures

In KIA, there are canals of different widths and the total length of the seawalls is nearly 70 km. Some of the seawalls are not used for industrial purposes or covered by green thus providing opportunities for redevelopment with high level of amenity [8]. Historic industrial structures such as shipyards, factories and warehouses can be found in some sites. These are valuable resources for urban regeneration.

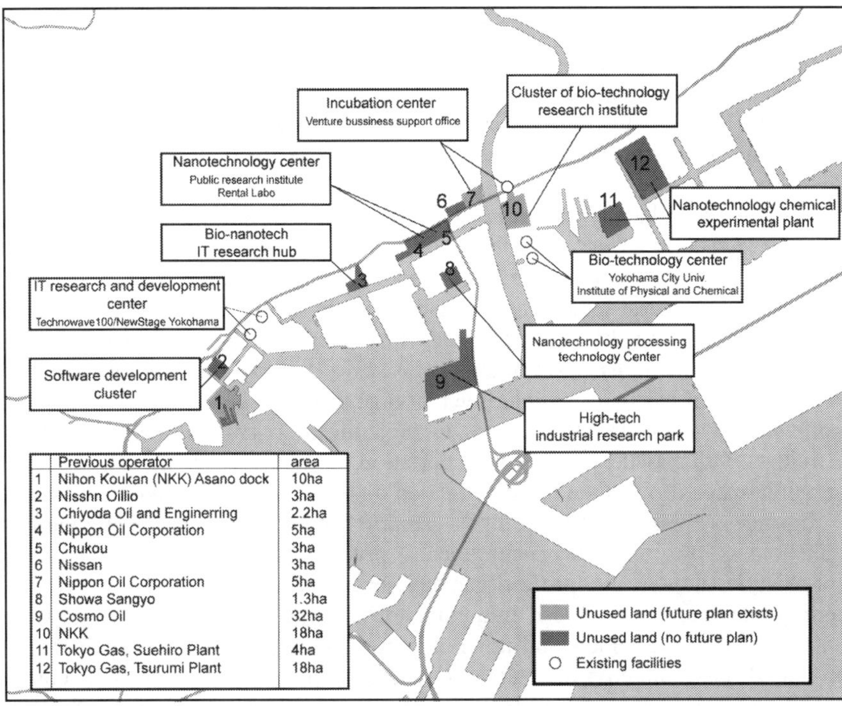

Figure 5: Underutilized sites and their proposed land use transition [7].

3.2.4 Environment

KIA has been a largest polluter to the regional environment. In recent years, more efforts have been called for to address the slow improvement on the regional environmental quality but a large improvement has been made in the past through the effort of industries responding to the gradual enforcement of national legislations. For instance, 80% of nitrogen oxides and 98% of sulphur dioxide emission loads to air were cut in the last 30–35 years [9]. On water environment, the First Total Pollution Load Control Scheme laid down by the national government in 1984 accelerated the reduction of water pollution loads [10, 11].

Regarding the soil contamination, accumulated survey results start revealing the situation in both Kawasaki City and Yokohama City, embracing KIA. In 2000, Kawasaki City reported 79 cases of site contamination [12] while 87 cases were reported in Yokohama City in 2004 [11]. This is, however, seen as a tip of icebergs and more cases are assumed to be present in KIA.

4 Conclusion: toward the integrated methodology

To develop a comprehensive plan for KIA, both top-down and bottom-up approaches are needed. Top-down approach involves the analysis of land use demand and industrial development potential from the regional perspective. However, this approach has its limitations since there is no organization in charge of conducting integrated regional planning for the whole Tokyo region. Each of the four prefectures has its own strategies for land use and industrial development thus making it difficult for Kanagawa Prefecture or City of Yokohama to direct industrial development KIA or estimate land use demand in KIA.

On the other hand, bottom-up approach involves site-by-site analyses of underutilized sites to estimate their redevelopment potentials and alternatives taking into account the degree of soil contamination, future land use, best-suited cleanup technologies, legal schemes and financial incentives possibly available for site remediation, historic and green resources, and location in relation to transportation facilities. An important key to starting this approach is the development of easily accessible inventory/database of contaminated properties, using public available information, to be utilized in any following planning activities. Other issues such as transportation, building types and urban design of public space can be then discussed based on the results of the above analyses.

In the process of developing a comprehensive plan for KIA, various stakeholders including city departments, property owners, business owners, potential developers, non-profit organizations and citizens should be involved in appropriate stages. It is also important to have the professionals support the process including planners and urban designers who generate alternatives based on the discussion, environmental and other experts who conduct scientific analysis and evaluate the alternatives as well as facilitators and managers who run the process efficiently and effectively.

References

[1] Ministry of Internal Affairs & Communications, *Population Census*, 2005

[2] Ministry of Economy, Trade & Industry, *National Census of Manufacture*, 2005

[3] Ministry of the Environment, *FY2003 Summary of Enactment Situation of the Soil Contamination Countermeasures Law and Survey Results on the Cases of Soil Countermeasures Assessment and Countermeasures,* 2005

[4] Geo-Environmental Protection Center, *FY2003 Survey Results on "Soil Contamination Assessment and Countermeasures",* 2003

[5] Kanagawa Pref., *Keihin Waterfront Area Regeneration Framework*, 1997

[6] City of Yokohama, *Keihin Waterfront Area Regeneration Master Plan*, 1997

[7] City of Yokohama, *Keihin Waterfront Area Urban Regeneration Basic Survey*, 2002

[8] Center for Sustainable Urban Regeneration, *The University of Tokyo, Keihin Waterfront Area Action Study 2004 Annual Report*, 2004

[9] City of Yokohama, *Yokohama White Paper on Environment 2004*, 2004

[10] City of Yokohama, *Environment in Yokohama: Annual Report on Environmental Management in Yokoyama 2005*, 2005

[11] City of Yokohama, *Yokohama Water Environment Plan*, 1994

[12] Mainichi News Paper article, *Increasing cases of site contamination in Kawasaki: shifting use of brownfield to housing*, June 23, 2002

From contaminated site to premier urban greenspace: investigating the success of Thames Barrier Park, London

J. Villella, G. Sellers, A. J. Moffat & T. R. Hutchings
Environmental and Human Sciences Division, Forest Research, UK

Abstract

Thames Barrier Park, opened in 2000, is situated on the River Thames adjacent to the Thames flood barrier in the ethnically diverse and economically deprived London borough of Newham (LBN). At nine hectares, it was one of Britain's most expensive areas of greenspace development costing £12 m to construct on a formerly contaminated wharf. With annual running costs of £700k, it far exceeds the budgets available for the maintenance of similar greenspaces in the area. It was designed by French landscape designers using Parc Citroen in Paris for inspiration. The park's original aims were: to be an area of multifunctional greenspace amongst new apartment and commercial developments, to improve surrounding house and land prices and promote the area as a pleasant place to live and work, to become a tourist destination and be seen as an integral part of a wider scheme that would draw tourists into the area. The aim of this paper is to assess whether the park was meeting or would meet its objectives. An independent survey of the local community and park users, utilising postal surveys, user questionnaires and focus groups was undertaken. Results suggested that the park had been as successful in attracting local and non-local users. Users of the park did not reflect the make up of the local population, tending to attract slightly wealthier members of the community who were predominantly white, in an area of high ethnicity and low income. This paper identifies future developments that will possibly transform this park's prospects of meeting its original aims and offers proposals on ways to achieve its wider objectives whilst meeting the needs of the local community and highlights important considerations for future urban greenspace design.
Keywords: contamination, brownfield regeneration greenspace urban park, tourism ethnicity, community.

WIT Transactions on Ecology and the Environment, Vol 94, © 2006 WIT Press
www.witpress.com, ISSN 1743-3541 (on-line)
doi:10.2495/BF060161

1 Introduction

Thames Barrier Park, fig 1, is constructed on a former dock and factory site that handled and processed industrial chemicals. After closure of the older docks in the 1960s, these contaminated sites became derelict. However, with the creation of the London Docklands Development Corporation (LDDC) in 1981 these areas began to be regenerated, mainly for housing and commercial development. A number of sites, of which Thames Barrier Park is just one, were converted to premier greenspace use.

Thames Barrier Park aimed to achieve a number of objectives. It was to be an area of multifunctional greenspace for the local population amongst new apartments and commercial developments. It was to improve surrounding land and house prices so that the area would be seen as an pleasant place to live, work and invest in, and it was to be an integral part of a projected wider tourist development plan, to include an aquarium development opposite the park, that would draw tourists into London Docklands, Riley [1].

The park is situated in the London Borough of Newham (LBN) adjacent to the River Thames in east London. Newham is an ethnically diverse borough due to its proximity to the old docklands and has a number of migrant communities LBN [2].

The creation of the park took place in two phases. The first phase began in 1997, and involved clearing the derelict land of physical structures and the removal of visibly contaminated soils together with a 300 mm layer of crushed concrete that had been added previously as a capping material. A 'cut and fill' operation created the final formation profile and a capillary break layer was added to seal off the subsoil beneath. Constructed areas such as a 'Green Dock' and fountain area were also created Riley [1]. Phase Two began in 1998 and involved the primary landscaping works. A mainly clayey topsoil was placed over the capillary break layer to a depth of between 1 m for the wildflower meadow and grass areas to 2.5 m for areas planted with trees. As part of the wider regeneration remit for the Newham and Docklands area, a Docklands Light Railway (DLR) station was constructed in December 2005 at the main park entrance, which should improve access to the park and bring in visitors from other areas of London. It is envisaged that this will help the park compete not only with other suburban parks at Greenwich and Battersea, but also the major city centre parks such as Green Park and St James' Park. Furthermore, the 2012 Olympics, awarded to London, will be predominantly based in Newham and Docklands, which should improve the visitor prospects for the park.

The park is 8.9 hectares in size and cost £12 million to construct. An annual budget of £700 000 per annum is used for park management, landscape maintenance, and marketing Lewis [3]. The objectives of this study were to investigate whether the park had, or was on the way to meeting, its stated objectives and suggest practical solutions to improve the functional usage of the park in the future.

2 Methodology

2.1 Community postal questionnaire

Five hundred community questionnaires were sent out to residents in the London Borough of Newham. The questionnaire covered subjects such as people's opinions of parks and what they considered an ideal park, whether they visited parks, in particular Thames Barrier Park, and what mode of transport they used to travel there. It included subjects such as age, ethnicity and income.

2.1.1 Thames Barrier Park on-site visitor survey

Two sets of information were collected. A count of the number of visitors entering the park was made and a visitor survey was conducted. The on-site survey took place over three days on Saturday 30th June (8.30 – 18.30), Sunday 31st June (8.30 - 18.30) and Tuesday 2nd August (8.30 – 18.30) 2005. It was in the form of a questionnaire where people either filled in the form directly or were asked the questions verbally and the sheet filled in by the canvasser. A total of 103 surveys were completed, the majority in the first two days. Towards the end, the people visiting the park tended to be repeat visitors who had been surveyed in the first two days. The survey sought to determine who used the park in terms of gender, age, income and ethnicity, how they used the park, how they travelled to the park, what they liked about the park and how the park could be improved.

2.2 Focus groups

Eight members of the local community were invited to attend a focus group session on the 19th July 2005. There were three women and five men from a variety of socio-economic backgrounds but they all came from a white ethnic background. The aim of the focus group was to allow the participants to explore further their feelings about the park that had been collectively raised in the questionnaire survey.

3 Results

3.1 Community questionnaire

Of the 500 questionnaires sent out, only 34 were returned (6.8%) which is a poor response and inevitably limits the strength of the conclusions that can be drawn about the greenspace values that are held by the inhabitants of Newham. Nevertheless, the results indicate that there are probably differences in attitude between those who use the park and those who are neighbours to it, which require study and a more narrowly focused sampling strategy.

Of the 34 park neighbour responses, only five had never visited the park and only two of the respondents were men. The reasons for not visiting the park were that they had never heard of it, they did not have enough time to visit, they could not access the park because of a disability or had not because of fear of intimidation.

As shown in table 1,the majority of respondents were either: people in their 30s or those aged 60+. The other age categories were not nearly as likely to use the park.

When it came to a question of income 44% did not answer the question. The census figures on Newham, LBN [2], say that the average household income is £27 000. Table 1 suggests that the community survey data does reflect the census data. However, as table 1 shows, he community survey did not reflect the census results for ethnicity, which identifies Newham as a multi racial community with a white population of 40%. In contrast, the community survey suggests that respondents were overwhelmingly of white ethnicity.

Table 1: Summary of age, income and ethnicity data obtained from the community survey.

Age	20 - 29	30 - 39	40 - 49	50 - 59	60+	
Percentage who replied	11.8	29.4	5.9	2.9	35.3	
Gross Household Income	£0 – 9 999	£10 000 - 19 999	£20 000 - 29 000	£30 000 - 39 999	£40 000 - 49 999	£50 000 - 99 999
Percentage who replied	22.2	11.1	7.4	0	0	3.7
Ethnicity	White	Black	Asian	Chinese	Mixed	
Percentage who replied	70.6	5.9	2.9	2.9	5.9	

3.2 Thames Barrier Park on-site visitor survey

One hundred and three completed surveys produced a representative sample (12%) of the total number of people (874, including those that visited more than once over the survey three days) that visited the park over the survey period.

3.2.1 Where visitors have travelled from

The results, fig 1, indicate that 57% of the visitors came from the local area of Newham and Tower Hamlets (the neighbouring borough) with 43% being attracted to the park from further afield.

3.2.2 Areas of the park used by visitors

The results, fig 2, suggest that most of the major areas of the park were used equally apart from the more specialist areas such as the basket ball court and the children's play fountains.

3.2.3 Park user activities

The main activity, fig 3, was taking the children out to play followed closely by relaxing. Exercise or dog walking formed relatively small proportion of the overall activity in the park.

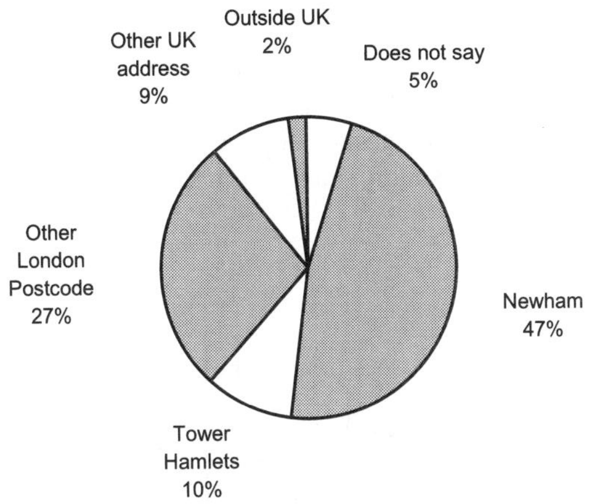

Figure 1: Where visitors have travelled from.

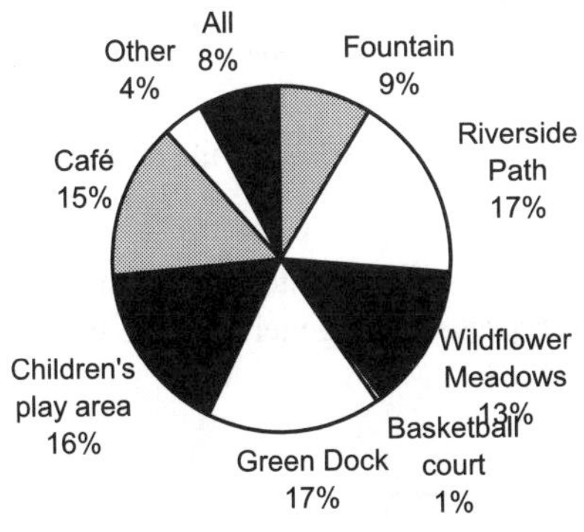

Figure 2: Areas of the park used by visitors.

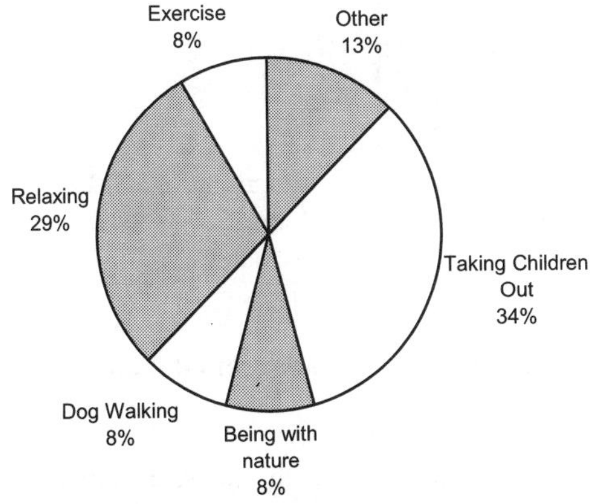

Figure 3: Park user activities.

3.2.4 Age and gender

The gender distribution, fig 4, favours female (59%) to male visitors (42%). When gender is broken down by age, the majority of users in their 30s and 40s were females while men made up the majority of people over the age of 60. Overall, the visitor survey suggests that people using the park are predominantly in their 30s and 40s. Interestingly, there was no recorded use by teenagers.

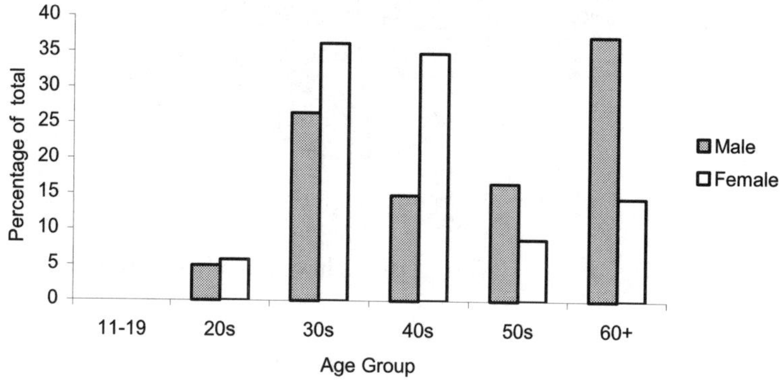

Figure 4: Park user age and gender.

3.2.5 Park user ethnicity

The majority of the park users were white, fig 5, with very few minorities using the park. This does not represent the general ethnic community mix in Newham of 40% white and 20% black.

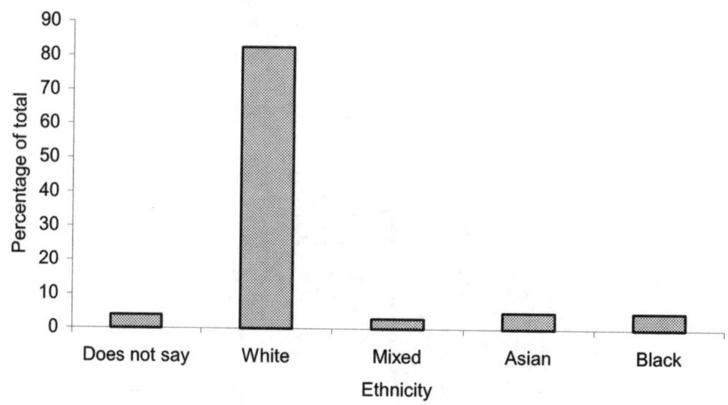

Figure 5: Park user ethnicity.

3.2.6 Income

Forty seven percent of respondents left this question unanswered. Of those that did, fig 6 shows that most had an annual income above Newham's average of £27 000 (LBN [2]).

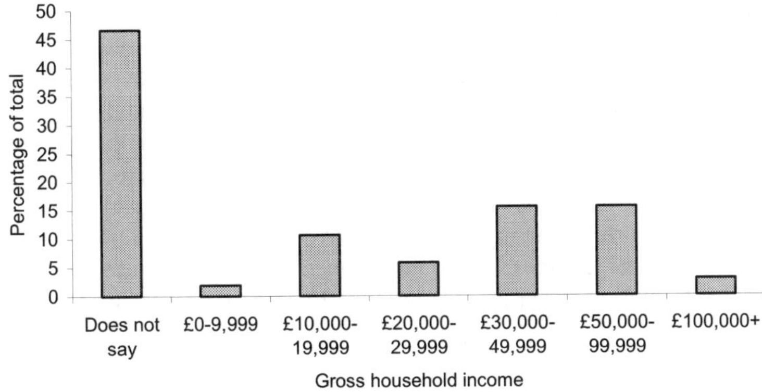

Figure 6: Park user gross household income.

3.2.7 Transport to the park

The results show, fig 7, that the vast majority of people travelled to the park by car. Very few came by public transport, on foot or cycled. These findings bear out those on income above.

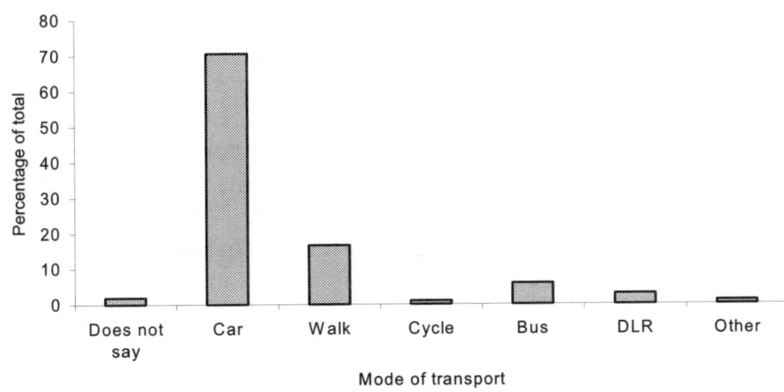

Figure 7: Mode of transport to the park

3.3 Focus group results

The focus group agreed that Thames Barrier Park was close to their idea of an 'ideal park', which they defined as being suitable for children, safe, clean and relaxing. Many felt that it was underused and some referred to the park as being 'like my own secret garden'. Although there was a security guard present, people in the early mornings and evenings felt the lack of people made them insecure. They commented that the size of the open spaces and wildflower areas meant that teenage people did not use it to play ball games, which appealed to the older visitors and those with young children.

There were some negative comments about the park. Those surveyed felt that the imminent removal of the daytime park security guard, would make the park less safe and much more like the local council parks nearby, and threaten future use by a significant group of current users of the park. People were divided on the need for wildflower areas, some seeing them as scruffy and others valuing their ecological value and aesthetic appeal. However, all users thought that they would understand the importance and function of such areas if information boards were erected. This was a comment towards the park with a user consensus that more information should be provided describing the park in terms of the habitats it provides for plant species and communities, wild fauna, historical information about the site, and the history and function of the Thames Barrier.

4 Discussion and conclusion

The first objective of the regeneration was that the park was to become a multifunctional space for the local community. Although the results from the community survey were disappointing and the data too weak to form reliable conclusions from, the visitor survey and the focus group results suggest that the park has been successful at meeting this objective. The majority of visitors were from the local community, fig 1, and they appreciated and visited most of the features that the park offers, fig 2, undertaking a wide range of activities fig 3. However certain age groups, such as teenagers, do not use the park fig 4, possibly because there is insufficient open space to play ball games. In comparison to the other age groups, there were fewer people between the ages of 40 and 50 and yet, at 23.6%, they make up the second largest age group in Newham, (LBN [2]). Results suggest that the mainly female 30-40 age group are accompanying younger children to play in the park, as figure 3 shows that the majority of people use the park to entertain children. The second largest group are visiting for non-physical relaxation with few taking part in exercise. This is almost certainly down to the restricted size of the park and may be a deterrent to other more active age groups or sectors of society.

There are some objectives that the park has been less successful in achieving. The focus group thought that it was a pleasant piece of greenspace which added value to the community in terms of improving the area as a place to live, but the survey results suggest that the park is not reaching out to all members of the community. The ethnic mix of people visiting the park, fig 5, does not reflect the census results (LBN [2]). It wasn't clear from the focus group or community survey why ethnic groups don't visit the park but ethnic exclusion has been highlighted in many studies and reports [4, 5]. The results also suggest that the park is predominantly attracting higher income groups than is representative of the area, fig 6, according to the census data LBN [2]. This may be because the housing immediately surrounding the park are expensive apartment dwellings and proportionately more people from these apartments are using the park compared to people from areas with lower average income. A second reason may be that the park is separated from other Newham communities by a main road, with very large traffic density, and this makes access to the park less easy for people without a car. The census data for Newham shows that almost half the households in Newham, 48.9% have no vehicular transport at all LBN [2]. Compared to the census data, the park is disproportionately attracting those who own a car, which may be a reflection of those who have the financial means to own one.

Another objective of the park was to develop a destination attraction in its own right and form a significant contribution to the wider tourist development plan for Docklands. Forty-three percent, fig 1, of the visitors to the park came from outside the local area which, given the park's situation and relative newness, should be considered a success.

The park has therefore been successful in creating a formal greenspace on former brownfield land that is used, appreciated and enjoyed by the local community and has had success in attracting some people from outside the local area. It has not been as successful in reaching out to ethnic members of the community or the lower income groups who are possibly excluded by the difficulties in visiting the park without car transport. However, the completion of Pontoon Dock DLR station in December 2005 should greatly enhance access to the site Lewis [6]. Furthermore, the imminent development of Silvertown Quay directly opposite the park, which includes a new aquarium and walkways over the main road should enhance the attraction of the park and its pedestrian access Lewis [6]. Moreover, the 2012 London Olympics to be held in the area, may also draw people to the park. So the future potential for Thames Barrier Park to meet its objectives for the local community and its wider aims to be a destination site is considerable. However, emphasis may be needed to develop programmes and strategies to attract people from ethnic communities and lower incomes.

Results from Thames Barrier have wider reaching implications for future provision of urban greenspace. Many regeneration projects for green space require them to be age inclusive. Our study suggests that this may not be possible in reality nor be desired by important sectors of the community. Most greenspace regeneration projects are designed to maximise access to all ethnic groups. Our study shows that premier greenspace developments in areas with large numbers of people from differing ethnic backgrounds still fail to attract people from non white ethnic backgrounds.

Acknowledgements

We thank the SUBRIM consortium for financially supporting this work through funding from EPSRC. We also thank Paul Lewis and Phil Riley for the help, advice, information and time that they gave to this project.

References

[1] Riley, P., Personal communication, September 2004, Marketing and Events Manager, Thames Barrier Park, London, UK.

[2] London Borough of Newham, (LBN) 2005, census data. http://www.newham.gov.uk/content/democracy/census.jsp
[3] Lewis, P. Personal communication, March 2004, Park Manager, Thames Barrier Park, London, UK.
[4] Morris, N., Black and Minority Ethnic Groups and Public Open Space Literature Review, Edinburgh College of Art and Heriot-Watt University, Edinburgh, UK., 2003. http://www.openspace.eca.ac.uk/pdf/BlackMinorityLitRev.pdf
[5] Dunnett, N., Swanwick, C. & Woolley, H., Improving Urban Parks, Play Areas and Green Spaces, Office of the Deputy Prime Minister, London, UK, 2002. www.odpm.gov.uk/index.asp?id=1127727
[6] Lewis, P. Personal Communication, 9 February 2006, Park Manager, Thames Barrier, London, UK.

Learning from experience: creating sustainable urban greenspaces from brownfield sites

G. Sellers, T. R. Hutchings & A. J. Moffat
Environmental and Human Science Division, Forest Research, UK

Abstract

Russia Dock Woodland Park and Bow Creek Ecology Park (formerly Limmo Peninsular Ecology Park) were created by the London Docklands Development Corporation (LDDC) on brownfield sites in London. The original designs failed in part (Russia Dock) or in full (Bow Creek) to be sustainable. This paper reviews their history and development and highlights the lessons that could be learned. Bow Creek Ecology Park (Grid Reference: TQ391812) is a 2.5 ha greenspace in London Docklands. The design was ambitious consisting of six separate ecology zones from chalk grassland to fenland maintained by a complex sluice and wind powered water pumping system. Russia Dock Park (20 ha) is in Southwark (Grid Reference: TQ361798) where derelict wharf areas were utilised to create a wetland ecosystem surrounded by fast growing woodland. Water levels and flow were maintained by a sophisticated pumping system. The main failings at Bow Creek were that the wind pump system could not maintain the wetlands. Habitat areas were also small, so succession became difficult to prevent and the habitats began to merge. This was compounded by the lack of long-term funding available to maintain this restoration, or staff with the skills to run such a complex design. Similarly at Russia Dock, the pumping system, and thus the wetland ecosystem, failed and a general decline followed. In 2002, an improvement programme redeveloped the park on a more sustainable scale, improving the woodland planting scheme and simplifying the original water habitats. This study demonstrates that public consultation, the creation of low maintenance water pumping systems, and the provision of long-term aftercare programmes made during the planning stages of development can make a considerable difference to the sustainability of new greenspace containing wetland features.

Keywords: sustainable, greenspace, brownfield, woodland, ecology, Bow Creek, Russia Dock, wetland, ecology park.

WIT Transactions on Ecology and the Environment, Vol 94, © 2006 WIT Press
www.witpress.com, ISSN 1743-3541 (on-line)
doi:10.2495/BF060161

1 Introduction

The overall aim of this case study was to investigate the history of two brownfield sites that were turned into greenspace and parkland in the second half of the twentieth century by the London Docklands Development Agency (LDDC).

Both parks were expensively developed with complex landscape design plans. Both failed to live up to their original design ideals, fell into decline and had to be redeveloped again in this decade. In neither case, as far as can be ascertained, was there any community consultation as part of the original development planning process. Subsequently at both sites, a second redevelopment, following community consultations was undertaken. At Russia Dock, this led to a design that improved the woodland and open space areas of the park while simplifying the wetland and pond areas. At Bow Creek, which eventually had to be closed for health and safety reasons, the complex water features have been removed and replaced with seasonal wetlands and a pond which do not require pumping. The general layout of the park was simplified, with expensive raised decked walkways removed and normal tarmac paths put in their place. Both these redesigns should make the parks more sustainable in the longer term because they are less expensive and complex to maintain, have fewer mechanical features, such as pumps and wind turbines which malfunction, and are able to be run by rangers rather than by more specialist trained personnel.

Bow Creek was developed as an 'Ecology Park'. This concept originated in the Netherlands in the 1950s and represents the creation of several different habitats in an urban area so that urban inhabitants can experience and learn from them without having to go to the countryside and visit them individually. Their primary aim is therefore educational, especially for urban schools and colleges [1]. However, maintaining numbers of artificial habitats by preventing natural succession occurring and the preservation of artificial habitats within small areas can be expensive and labour intensive. The smaller the artificial habitats become, the more expensive and difficult this is [2]. Although Bow Creek was defined as an Ecology Park, there was never the long-term financial or professional resources put in place to maintain it adequately [1]. Furthermore, the poor design and quality of construction materials used hastened its demise as an educational resource.

Russia Dock is located in Southwark and is part of the Surrey Docks redevelopment that was begun by the LDDC in the late 1970s. It is a linear greenspace development of around 20 ha, (fig. 1), which was formerly the Grand Surrey Inner Docks [3]. Some of the park is densely wooded with trees planted in the late 1970s and early 1980s to a 'naturalistic' landscape design. The main species were fast growing trees such as alder, willow and poplar, supplemented with more recent plantings of slower growing climax trees such as oak and beech. Other parts of the park open out into grass areas and the main paths follow the old wharf edges. Most of the dock was filled in but some was left to form water features, such as ponds, with interconnecting water ducts and streams. Water levels and flow were maintained by a complex water pumping system, but for a number of years most of these areas have been left to silt up so that they now form areas that are only wet in the winter [4]. There have been no recorded contamination issues associated with this site.

Figure 1: Aerial photograph of Russia Dock Park.

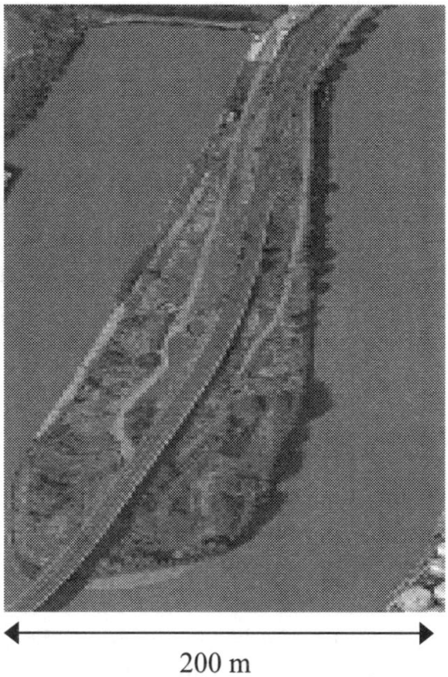

Figure 2: Aerial photograph of Bow Creek Park.

Bow Creek, a former railway yard, fig 2, was developed in 1996 when it was named the Limmo Peninsular Ecology Park. There were contaminated soils on the site but these were covered by a 200 mm layer of imported topsoil in 1991 as part of the development of the Docklands Light Railway (DLR) flyover, which bisects the park. Later reports commissioned by the LDDC [6] suggested that the land on the west side of the flyover could contain slightly elevated levels of mercury, so an extra 400 mm of topsoil was added there during park construction. The park was created with six small ecosystems which were designed as an educational tool for local schools and groups as well as being a valuable, if small, piece of greenspace in an area where it is lacking.

Both parks have had problems associated with them. By reviewing their history and analysing the problems that occurred, lessons on how to avoid similar situations in the future can be drawn.

2 Russia Dock

2.1 Developmental history

Russia Dock consists of the area formerly known as the Grand Surrey Inner Docks which were constructed from 1696. Previously, the land behind the old village of Rothehythe had been wet marshland unsuitable for farming. Rothehythe had become a centre for shipbuilding and the new docks, including the area now known as Russia Dock, became an extension of this. By the 19th century, the docks were handling grain, timber, cheese and bacon trades together with being a base for the Arctic whaling fleet [6]. As ships got bigger the docks became less capable of handling them and the docks were closed in 1969. In 1978, the LDDC filled in the docks with unconsolidated ex-construction material. Most of the area was designated for housing and commercial development but Russia Dock was restored as an area of greenspace and handed over to Southwark Council in 1982 [7].

2.2 1980–2002

Originally the park was to have been designed as a formal London Park but this concept was abandoned in favour of a less formal open space and woodland scheme [7]. Although there was no soil on site, there were quantities of silt from the bottom of the docks, which was removed and dried out. This was mixed with imported topsoil to create the growing medium for the park [7]. The park was laid out in blocks of intensively planted fast growing woodland punctuated by pathways connecting the adjoining housing areas. At its centre, the park opens out into an amenity greenspace, fig 1 [3]. Incorporated into the design was a complex system of streams, ponds and created wetlands which were to be maintained by pumps [4]. However, by 1986 the park and especially the water features had declined. A number of problems had become apparent and as early as 1986, just four years after completion, P. Jackson [8] commented:

" the site [Russia Dock]was handed back to the Borough [London Borough of Southwark] for maintenance. At this stage, the problem of expertise and resources required to manage the water features became apparent having never been addressed at the design stage. Following the drowning of a child, there was an unwillingness to make the water feature work. The full implications of managing a man made water feature, and maintaining water quality and circulation, became very critical. The result is that the problems of the water feature remain unsolved and a scar on the landscape."

After the drowning, a management decision was made to allow the water feature to dry up in places in order to lower the water level [8]. M. Pearson [9] makes the further comment that by 1999:

"...the interesting situation in 1986 has continued to manifest to the point whereby the tragic event of the drowning of a child still determines the strategy today for the maintenance of the water feature. And as a result of a non intervention strategy the landscape of the water feature and that which surrounds it has been allowed to deteriorate."

This shows us that, even as early as 1986, the park was in decline, the complexity of running the water feature had become critical and expertise to maintain it was not present. The situation continued to decline until 2002.

2.3 2002 to present day: sustainable *re*development

In 2002, £350 000 was assigned to the redevelopment of Russia Dock after years of decline and undermanagement [4]. A report on the park and a proposed management plan were commissioned from Voelker Consultants [3]. They identified a number of areas of concern. For example, the open water system had been vandalised, had silted up and fallen into disrepair leading to stagnant surface water. After years of underfunding, the general infrastructure of the park had also fallen into disrepair, with pathways overgrown and bridges, lighting and railings falling into decay. The three timber bridges crossing the water bodies needed to be replaced and the two steel bridges were in need of refurbishment.

As a result of this report, a major redevelopment programme was initiated. Around £40 000 was spent on an urgent tree thinning programme to remove the decaying poplars and willow which had become dangerous and spindly because they had never been thinned out in the 1980s. A programme was instituted to gradually replace poplars with slower growing climax species such as beech and oaks, which were not specified in the original design specification [4].

The process of allowing most of the water features to green over was continued and they have duly become seasonal which improves their conservation value. However, there has been some repair and redevelopment of a few of the ponds and water conduits. Water is now pumped between them, which has allowed the return of fish to some of the ponds. Pond redevelopment is on a much more manageable scale and is far less complex than the original installation and therefore easier to maintain physically and financially [4].

New paths were put in and the old paths restored to make access and movement around the park easier. Infrastructure was also improved with more seats and better lighting [4]. The new, more robust infrastructure, the thinning of trees and the replacement of the poplar trees, together with the partially restored, simplified water system should make the park sustainable in the longer term.

2.4 Problems identified in the park

The history of this park shows that a number of identifiable problems arose almost immediately upon the park's opening. This was not a park that gradually declined over time. It declined immediately and critically.

The main problem was highlighted within four years of the park opening by Jackson [8]. The design stage never addressed the amount of expertise and resources that would be needed to maintain the park, especially the complex water features. The situation was further compounded by the death of a child soon after opening, which led to a decision not to maintain the water features [7]. As Pearson [9] points out, the death seemed, until at least 1999, to be the key factor that determined the non-intervention strategy which was adopted with regard to the water features.

The lack of sufficient expertise and resources impacted on other areas of the park. Trees were not thinned out when they should have been, so they became spindly and eventually dangerous, particularly so with the poplars. This led to an emergency remedial thinning and replacement programme in 2002.

The minimal maintenance programme seemed to be restricted to cutting the grass and clearing up litter [8] which meant that the wildlife value of the park suffered, and vandalism and quad bikes became a problem [4].

Many of these problems could probably have been identified at the planning stage, especially the resources and expertise needed to maintain the park, but it seems they were not. Furthermore, there was little evidence of any community consultation, which would undoubtedly have highlighted potential community concerns, such as vandalism and arson.

None of these problems were addressed for over 16 years until money became available to carry out a substantial remedial programme to correct the many potentially avoidable problems found in the park.

3 Bow Creek Ecology Park

3.1 Developmental history

Historical information is limited for the Bow Creek site. In 1805 the Limmo Peninsular was still wet marshland, but in the Victorian period it became a coal yard and a coal wharf for the Thames Ironworks and shipping company. This closed in 1912 and for 68 years Bow Creek became part of a railway marshalling yard [9]. On closure of the marshalling yards in 1970 the site remained derelict and unused until it was taken over, first by the Docklands Light Railway in 1991 to construct the line which cuts through the centre of the site, and then by the LDDC in 1994 who decided to develop the land as an Ecology Park and green space. When the LDDC was dissolved in 1998, the park was handed over to the Lee Valley Regional Park Authority (LVRPA) [9].

3.2 Park development 1993–1998

In 1993, the LDDC held a design competition which was won by Gibberd Landscape Design. The design philosophy was to utilise the already existing freshwater ponds that had naturally developed and create six different habitats. These would be connected via gravel paths and non-slip decking [10]. The new Ecology Park was to be primarily developed as an educational resource for school children so that they could see a number of different habitats within an urban area lacking in greenspace. The six habitats chosen were chalk grassland, willow coppice, mixed scrub, fen, ditches and ponds [10]. The wetland and water features were to be fed by a wind pump that pumped water into a tank and then through the park and habitats via constructed water conduits and sluices [10]. Construction began in 1994 and was completed in 1996. The park was managed by the Trust for Urban Ecology until 2002 [1].

3.3 Park decline 1998–2000

After the LVRPA took over responsibility for the park it was realised that there were a number of critical problems which resulted in closure of the park for health and safety reasons. As early as 1996, problems with water supply to the water features had been recognised. Dewhurst and Wurzell noted in their habitat survey report [11], that the newly installed wind pump to supply the water features had failed and that the park was having to be supplied by the river and rainwater runoff from the DLR viaduct. However, this was inadequate and the areas described as the pond and chalk channel, together with the willow coppice and water meadow were not being supplied with sufficient water and consequently were not developing as habitats in the way that was originally envisaged.

The 2002 LVRPA Management Plan [13] identified a number of critical problems that had arisen since 1998 leading to eventual closure. These were:

3.3.1 Site infrastructure

The Management Plan [13] highlighted design flaws and maintenance failings. Many of the channels used to move water around the park were subsiding through ‘poor design’. The river edge fencing made of chestnut paling was prone to vandalism and did not reach health and safety standards. There was also a water wheel that served no purpose. The site infrastructure had deteriorated through bad design and maintenance and any site improvement programme ‘would have to address its purpose, sustainability and durability issues’.

3.3.2 Access

The main access into and through the site was a gravel path with boardwalks connecting to the ecological zones. The gravel path was found to be suffering from encroachment by weeds and many of the boardwalks had deteriorated, were slippery or had been laid at the wrong angle. Furthermore, the access routes did not meet the requirements of the Disability Discrimination Act.

3.3.3 Habitats and communities

The report commented that:

‘Due to the small size of the individual habitats, there is a problem with succession. Wetland habitats are also reliant on an artificial water supply. Poor design of the site has highlighted the fragility of the infrastructure in general, whilst the west shore suffers from tidal erosion. The site is affected by pollution through traffic congestion, factory emissions and water contamination. The site is also isolated and therefore vulnerable to vandalism.’

3.3.4 Public interest

The report noted that while currently there was no public access because of the construction work at the entrance to the park, the future was more promising because of the comparative dearth of greenspace and open spaces for public use in the area. However, it noted that the park suffered in the past because there was never any provision for visitor facilities or interpretation boards.

3.3.5 Education facilities

The report found that the health and safety issues identified in two Risk Assessment reports in 1998 and 2002 meant that the park was inadequate by LVRPA standards for educational programmes. This was mainly because the site was isolated and difficult to access.

3.4 The new development strategy for the park

Following critical design and maintenance flaws that were identified in the Bow Creek management plan [13], it was decided that the park would need renovating and simplifying. This is an ongoing programme begun in 2003. Essentially, the water features have been simplified because the water is no longer pumped, so there is now only one pond with other wetland areas becoming naturally wet only in the winter. The sluice systems, waterwheel and wind pump have also been removed. The majority of the boardwalking was also removed except around the dipping pond. The old pathways have been removed and one main tarmac path, forming a loop around the two halves of the park, has been created to replace the several smaller gravel paths. Seating and fencing has been improved and is vandal resistant. Attempts to maintain six small complex ecosystems have been abandoned and all but one of the former pond areas have been allowed to develop along more natural lines. There has been some limited tree planting and some areas are cut as part of a wildflower meadow maintenance programme. There has also been some planting of marsh species in the flood hollows.

The aim of the new management regime has been to correct the flaws in the original design by either replacement or removal. Through adoption of a modest design and simplification of the park it is hoped that it will now be both financially and physically more sustainable.

4 Discussion and conclusions

There are a number of similarities between the factors leading to failings at the two parks. Firstly, no allowance was made at the design planning stage for financing the long-term management of the parks other than to hand them over to new management agencies who had had little influence over their design. In both cases, the LDDC chose very ambitious designs for the parks, which at best, would be very demanding and expensive to maintain. However, the parks were constructed without any analysis of actual resources available to maintain the parks once they were handed over [1, 7]. The result was that the budget available to maintain each park was inadequate and the consequences were an almost immediate decline in park quality. The second similarity was that there was little public consultation over the original park designs. Had there been, then it is possible that less complex but more ecologically and financially sustainable designs may have been chosen at the beginning. The third similarity is that in order to be more sustainable, the parks needed to be regenerated and radically simplified. It would appear that there was confusion in the original design regarding the relative importance of naturalistic and artificial elements, and the interplay between them. There also appeared a worthy yet impracticable reliance on so-called 'green' solutions, such as the use of wind pump, which can be extremely difficult to manage and maintain in an urban setting. Ecology appears increasingly to be an important driver in urban greenspace design [13] but to be sustainable, areas designated within new urban greenspace projects must adhere to sound ecological principles.

In addition, deliberate construction of habitats ecologically alien to the surroundings will inevitably lead to larger maintenance demands. In this context, the ambitions of new 'Ecology Parks' must be tailored to the expectations of what resources will be available for maintenance. Educational extension services will also be necessary to gain the benefits implicit in Ecology Park construction. Public consultation will be essential to ensure that ecology is an important design driver for the new greenspace, as well as identifying the types of preferred habitats, and whether there is local expertise to support them, e.g. in a voluntary capacity. However, Russia Dock and Bow Creek are now success stories because eventually the problems identified above have been overcome through regeneration and redesign plans which included public consultation. The simplifying of the parks' aims and designs should mean that they are sustainable in the long term and will be a considerable greenspace asset to the surrounding communities for years to come.

The lessons to be learnt from these parks are that there should be a long-term financial and management strategy agreed at the time of the park planning and design stage so that the long-term sustainability of the parks, both financially and ecologically, can be assured. This should be achieved through public consultation, taking on board the views of the public, experts such as ecologists and the organisations that will need to maintain the parks post handover. It is only through doing this that a truly sustainable park design will result that will be an asset to the community over the long term.

Acknowledgements

We thank the EPSRC, who funded this research work as part of the SUBR:IM consortium under Grant GR/S148809/01. We also thank Gez Breed at the LVRPA and Jon Best at the London Borough of Southwark for providing their help, advice, information and time.

References

[1] Dewhurst, B. Personal communication, 17 March 2006, Chief Executive, Trust for Urban Ecology. UK.

[2] Watts, K. Personal communication ,16 March 2006, Landscape Ecologist, Forest Research UK.

[3] A Management Strategy for Russia Dock Woodlands, Voelcker Consultants, London UK, 2002.

[4] Best, J. Personal communication 26 April 2005, Ecologist, London Borough of Southwark, London UK.

[5] An investigation of the acceptability of the education and recreational use of the Limmo Nature Park by children, Clayton Environmental Consultants Ltd. Ilford, UK, 1994.

[6] Surrey Docks, LDDC Completion Booklets; London Docklands Development Corporation

[7] www.lddc-history.org.uk/surrey/index.html#History

[8] Jackson, P. Landscape as a catalyst. Landscape Design, 163 (October), pp 38-41, 1986

[9] Pearson, M. A landscape assessment: Russia Dock water feature. Thesis, Advanced Diploma in Arboriculture and Community Forest Management, Middlesex University, 1999.

[10] Breed, G. Personal communication 15 April 2004 Ranger Lee Valley Parks Authority. London UK.

[11] Landscape developments at Limmo Peninsular, London Docklands Development Corporation, 1995.

[12] Bow Creek Ecology Park, The Bird Sanctuary and East India Dock Basin habitat survey, London Docklands Development Corporation, 1996.

[13] Bow Creek Management Plan 2002 –2005, Lee Valley Park Authority, 2002.

[14] Crystal Palace Park. Public Consultation. The report, the results , 2002.

Section 6
Community and public involvement

Brownfields – public involvement

A. Bogen
President, Down To Earth Consulting, LLC, USA

Abstract

It is the public's money that provides critical leverage in Brownfields revitalization. Without grants for environmental assessments or cleanup funding or tax incentives, many Brownfields sites would remain unkempt. Municipal staffs are public employees whose actions in the Brownfields process must be transparent. Regulators too must share findings and answer an enormous range of technically challenging health and ecological questions. The press seems to find stories of environmental resolution worthy copy. Their questions could be difficult to manage when the property owner may be wrestling with the responsibilities and liabilities of potential site contamination. Sometimes, in neighbourhood meetings of directly effected parties, the whole process is characterized as being self serving for developers, bureaucratic, costly and time consuming and indifferent to the real concerns of the public. The charge for all parties involved in Brownfields redevelopment is to foster the public good by resolving ecological and human health risks and fostering economic development to increase taxes and employment. A trusted and effective process can assist Brownfields resolution. Time delays from litigation can be minimized. Infrastructure improvements are more likely to receive favourable bonding votes. Sophisticated and able developers are more likely to stay committed. This paper will be about developing just such a favourable process.
Keywords: grants, leverage, bonding, increased taxes, process, trust, liability.

1 Brownfields: developing a public process

The public bears the burden of Brownfields. They endure the health risks, suffer the loss of services not provided because of diminished tax rolls, and experience the deflating presence of blight. If the sites are to be returned to beneficial reuse, public support is critical.

WIT Transactions on Ecology and the Environment, Vol 94, © 2006 WIT Press
www.witpress.com, ISSN 1743-3541 (on-line)
doi:10.2495/BF060171

There are a few essentials that should be in place before the public is invited into an active role. These essentials are a competent administrator, some funding and established relationships with regulators. It would be very useful if the administrator or consultant has Brownfields experience. Hopefully that person will have attained an awareness of the patience, technology and partnerships so necessary to success. The sponsoring governmental entity needs to have held a discussion of goals and expectations for the program. Some Brownfields can take a long time to resolve. At this stage of the international Brownfields experience, many of the less complex sites have been uncovered and reused. Much of what remains may have serious issues. Therefore the expectations of success must be in the long-term work. To most fully serve the community's long-term needs, the Brownfields program ought to coordinate with other land use groups like transportation, economic development, zoning and conservation.

The first step in developing public support is a commitment to an open process by the governmental entity in charge. An open process is defined as one that shares information and decision making. Trust has to be developed. It is the public's money that will be allocated to Brownfields. People will want to know who receives and benefits from that funding. There is a delicate balance to be struck among the intentions of transparency, effective management activities and the complexity and confidentiality of deal making. An effective committee can help foster community interest and sustain commitment over time.

1.1 Forming a committee

The first committee members will probably be people familiar to the administrator and the governing body. Diverse yet inclusive membership is a vital objective to pursue over time. There is value in having specific skills and knowledge among members.

As an example, health professionals will be able to help residents understand the health risks. There are often concerns over the potential for cancer clusters. A proposed reuse may involve increased vehicle traffic to a site with the concurrent concerns about air and noise pollution. The public will want to know about the plans for resolving contamination on the different sites. There may be very vocal persons who prefer an alternative that is not economically feasible or required by regulation. A professional can lend direction and help the group reach an outcome to the debate.

Economic development professionals will understand the barriers to reuse associated with assessment and cleanup costs. They can support the effort to develop information as needed rather than to expend a lot of funds studying everything at the start. They may be able to help arrange contacts with informed developers. Bankers and non-traditional lenders may be attracted to assist in funding the redevelopment because of the relationships with these professionals. They may also have access to government programs for other sources of development funds.

A grasp of technical issues can be paramount to the management of costs. Therefore it is important to have representation from some of the associated sciences. For example, chemists and geologists can have a working knowledge of the hazardous materials encountered on a site and of the potential for ecological impact. It could be valuable to engage an educator from an environmental discipline or a volunteer professional from an engineering or environmental sciences company.

Representatives of the community are also important to have on the committee for a number of reasons. They can help their neighbors become informed of what the issues are and how they are being addressed in an informal manner. This can lead to important support for a reuse project rather than uninformed, fear based opposition. It also seems appropriate to listen to the concerns of those most directly affected by the Brownfields site. They have borne the greatest share of the burden. They also need to be part of the process when the decision is undertaken to prioritize sites for assessment funding. That is usually the first water shed decision. How it is handled can enormously enhance or damage the credibility of the process.

On going education of the committee members is critical to the strength and effectiveness of the group and the project. Lawyers and developers are often willing to come to such a group and to explain what is required or what is needed in a successful transaction. It takes time to understand the concept of a shared burden in the remediation of a Brownfield. A task may be both remedial and developmental. An example is the combination of removal of contaminated soil that is also geo-technically unsuitable. It would be cost efficient to accomplish those objectives as one coordinated activity. If the parties were a developer and a seller and they agree to share the cost, they might both have a net savings. The committee members, if familiar with such examples explained by experienced people, may come to see this dynamic as part of reaching a deal rather than aiding and abetting a polluter. The seller may nave entered the chain of title and therefore assumed liability but may not have caused the contamination. A Brownfields transaction may be the most sophisticated project some members have encountered in their careers. It is helpful to provide them with these transactional alternatives to promote flexibility, patience and dialogue. It is equally important to attend to regulation. Therefore these alternative ideas need to be presented as hypothetical so as not to create the impression that the technical standards are subject to manipulation. There can be political pressure that is uncomfortable to withstand. The best defense is in this open continuing process. Because Brownfields can be complex, the developers of the project can appear to be spoiled children demanding that everyone accommodate them so that they get what they want. If the goal remains healing the land and creating public benefit through a transaction, all parties can achieve their objectives. No one can work through all the hoops themselves. The committee is a vital part of the team making Brownfields alive again.

1.2 Committee functions

Foremost among the defined charges that the committee can carry out is the development of the criteria for site selection. These criteria will be very important when there is scrutiny as to why one site was selected over another.

The explanation needs to be clear when answering inquiries from the public or the press.

Public funds are scarce and precious. Competition for them is intense. Increasingly, projects need measurable returns to be funded. Usually communities are faced with resolving more than one Brownfield site. Therefore, they want to be able to continue to access additional funds over a period of years. Their assessments need to support a desired outcome. An environmental assessment requires rather significant funding. The work product has a shelf life. Regulations change as do site conditions. A completed study could be obsolete before active interest in the site materializes. There are too many troubled sites to attempt to build a library of data. Therefore a significant component of the selection criteria could be whether or not the site has real potential to bring a return on the public investment. The major considerations about whether or not to award funding can be developed through the following three questions:

Does the site have kinetic potential?
Is there an acceptable proposal from a capable developer?
Does the proposed reuse generate taxes and jobs or other public benefit?

If the answers to these questions are presented to it, the committee can compare and evaluate competing projects. An application form can be designed that leads the project advocates to define their answers clearly. For example, to determine the kinetic potential of the site to be reused, a number of questions can be considered.

Is there a written access agreement? If the current owner has reservations about allowing access, the study may never take place. These reservations need time to be addressed. The owner may fear that the information uncovered will force him into a costly remedial situation for which he is financially unprepared or unwilling to encounter. Defining the risk to the owner may not be the way to achieve cooperation. Instead it may be important to understand the site in question in relationship to the balance of the owner's other assets. Perhaps the owner assumes that the remedial costs will make the property inverted in value and is willing to transfer it to the community for a reduction in liability. If there is a public development interest in the site, as opposed to a delinquent tax action, the municipality may want to negotiate a deal pending the resolution of the environmental conditions. The owner may reluctantly agree but never the less agree to access. There are instances when the proprietary generation does not want to leave the burden to the next generation and so may agree to access. Sometimes there are negotiations that take place in the context of a pending or actual regulatory action. The owner may come to see the access as a start toward resolution. In some instances, the owner may see the public process as reassuring. The owner does not have to decipher what the technologists are saying. The use of the public money means that it should have some standard of accountability. Then they may not have to put up any funds to get started on developing the answers. If the owners come to see the inevitability of the need for resolution, they will try to reach accord. The committee needs to be kept informed of the process underway all along the way that it is occurring. They have to be able to support the negotiator and understand what importance the components of an agreement have to the parties. It is almost a mantra that needs to be recalled for each project. These sites are Brownfields and therefore are often complicated to resolve.

Is there a known health risk that could delay the process until it is resolved? There are diverse health issues that could effect a site selection decision. Sometimes, the committee may have a direct appeal from a health department so provide assistance in investigating suspected conditions on a site. For example, children may become ill from exposure to contaminants left on an unsecured site. The committee may ask why the health department doesn't conduct the assessment. In many instances, the health departments are looking at the effects on human beings and need assistance from another enforcement entity to clarify what the environmental situation is. Working cooperatively may get a site resolution underway. Not all outcomes need to be tax and job driven. Risk abatement can be a valid criterion. Perhaps the public and committee will enter into an enhanced dialogue about the uses of the site in question once the sources of concern become public information. At minimum, there will be a dialogue about abatement. The charge of the committee remains assessments. The likelihood of a resolution is increased when the baseline data is known and the real risk assessed. Otherwise these sites may stay blighted and worrisome.

Is there a tax lien and does the municipality control that lien? There are a variety of regulations that government bodies have about the procedures required when there is a delinquent tax bill. The municipality needs to have a high performing tax roll and Brownfields often owe back taxes. In some cases, the liens may have been sold and the purchaser wants full reimbursement and expected return. The municipality may need to go to court to create a universal resolution to this issue. There is no use conducting an assessment on a site that remains financially inverted in part due to the amount of the back tax bill. In some cases, a higher legislative body may hold the authority to grant the tax relief critical to getting the site resolved. The committee may want to see such sites as second tier priorities to be addressed when the tax matter is clear.

Does the owner stand to be unduly enriched if it receives this benefit? Is that owner responsible for other problems on other sites? In Brownfields work, public funds are being used to address an environmental and therefore economic situation that will affect a private party. The public perception of the benefit is critical to the ongoing health of the project. The law may hold the owner as the responsible party even if she did not cause the contamination. Assisting that innocent owner should still be a decision reached in dialogue. People will want to know that public funds are not being used as part of a bottom feeder development strategy. The committee needs to be careful that it is not manipulated into making a grant. It is also important to recall that the public's need for such grant leverage arose from an understanding that without it sites could remain blighted indefinitely.

There can be a wide divergence in opinions as to the acceptability of a development proposal. A projected increase in tax receipts may be attractive to the majority of the community but irrelevant to the adjacent neighborhood. The increased volume of traffic and concurrent risk to pedestrian safety and associated decline in air quality can be unacceptable trade offs. The community may also object to the specific reuse. The proposal could contain new forms of visual and noise pollution. Some times the existing infrastructure can't support the new project without additional public investment, which may or may not be acceptable. Therefore, it can greatly facilitate the work of the committee if the municipality has discussed these issues and provided guidance. It may be necessary for the committee to take the lead in getting the issues discussed.

The committee can also request additional information such as awareness of sustainable environmental practices in site configuration and building design. This can be a dialogue that educates and involves people. There is no need to defend regulations as the committee's primary charge – those tasks fall to the already established appropriate bodies. There is a growing awareness that reusing sites in a way that adds to off site runoff or significantly increases sewer use may add other unacceptable and costly ecological burdens. Reuse at any cost is contradictory to the intent of Brownfields resolution. Building with green orientation means support of sustainable reuse. The committee members will need continuing education to have this information. Brownfields efforts in this area can provide leadership in the general effort to build in a sustainable fashion.

1.3 Contractor selection

First among the committee functions related to their fiduciary charge involves the selection of contractors to perform the assessment. The administrator can prepare an advertisement for a Request for Qualifications. Attention will have to be paid to the regulatory requirements of such advertising. There is usually great response to these solicitations. The administrator may want to collate some information about the various companies to assist the committee in its review. For example, it may be useful to have comparisons of insurance capacity, technical staff qualifications, number of project managers, related in house technical and computer capacity, rates for services and lab fees and the like.

The committee will need to define what qualities and capacities it most wants from the contractor. There are firms specializing in certain remedial technologies and firms with more general practice. It can be inefficient to advertise and interview for each project. The committee may choose to establish a short list of contractors that stay eligible to bid on projects as these mature. There may be certain basic selection criteria such as having the appropriate licenses, insurance and bonding capacity. There are other qualifications that may be discovered in interviews where the contractors can discuss other projects that illustrate how they handled the myriad issues that often accompany a Brownfields assessment. The contractor's relationships with regulators can be explored. It may be possible to retrieve information from the regulator about the good standing of the license holders. The contractor can be asked directly whether any of their projects has matters in conflict. The committee can get a sense of how well the technicians can communicate to the laypeople on the committee and perhaps by inference to the neighborhood residents.

When the committee actually has a site assessment project that they put out for proposals, other factors emerge. They will need a basis to compare the different proposals. The contractors often provide a narrative and general cost estimate. It is suggested that the administrator limit the length of the narrative response. The technical skill does not expand in direct relationship to the volume of words used to describe corporate capacity.

The charge of the committee includes the circumspect use of the funds. They need specific budget lines to compare. They need to compare what tests are proposed, the number of tests, the cost of narrative reports and the like. They ought not to be required to always choose the low bid. There is a balance in every assessment among art, pragmatism and science. The training of technicians often leads them to suggest more sampling to be conservative and address the liability concerns of their insurance carriers. Part of the Brownfields assessment process can be seen as establishing the range of environmental concerns and the associated remedial cost rather than explicit detail on a site. The earlier stages of the assessment could serve to define the feasibility of the proposed project site to be remediated rather than full, specific details of a remediation plan. The first generation of data can usually provide enough information to determine that project feasibility. The issues could reveal extreme soil and/or ground water contamination. That cost of remediation could exceed the market value of the site. The municipality or other governing entity could still decide to allocate the necessary funding to create the leverage to restore the site.

The committee can also discuss rates for a particular contractor service to understand the range and value of those fees. Some governing units have arrangements with testing laboratories for lower rates, which these public projects may be able to access. A competent administrator is likely to understand the relevance to the developer and community of what is being proposed by the contractor. That person could also have a dialogue with the regulators to determine the necessity of some specific proposed activities. This will help insure obtaining the most value for the public investment.

Because the committee will independently select the contractors, the administrative staff is insulated from the pressures brought by the contractors' sales efforts. Separating the procurement and selection processes helps to reinforce integrity in contractor selection.

1.4 Project development

Soliciting sites is an ongoing activity. It is important for the administrator to establish and maintain contacts with all the parties who come into play with a real estate transaction. Developers, real estate agents and corporations often include the web in their search activities. Brownfields facilitators need to prepare internet information on sites they are managing to provide exposure for those sites and to foster more transparency. This and other more traditional networking venues can provide the opportunity for re-development options for the municipality. The developers in turn can receive information about the capacity of the Brownfields project to provide financial resources and support to developers. This interaction will also assist the administrator in her efforts to keep up with changing regulatory responses, new forms of funding and tax concessions and unique technical solutions. The field is very dynamic and in need of ever new approaches to financing. The resolution of the liability maze is an ever evolving process.

The detailed work of the committee begins when a site is recommended for an assessment. Staff needs to collate information about a site before it is presented to the committee for decision. The specific data that went into the site selection criteria grid needs to be recapped.

Some of the information may require sensitive and confidential management. For example, as discussed above, it may be that the site owner feels vulnerable to the process. She or he may have debt with the site pledged as collateral. If the environmental conditions reveal a costly clean up, the value of the site could become inverted. This may have an immediate and deleterious effect on the financial condition of the owner. Public corporations in the US must now make representations about those liabilities. Small owners and their lenders will likely have varying degrees of awareness of the situation. However, as mentioned above, staff and the process must take into account the financial repercussions of the proposed project. In the past, where there were just such concerns, owners have sometimes abandoned the sites for fear of the cost of remediation. It is critical to discuss the risks associated with granting permission for an assessment. The administrator needs access to and the support of lawyers with transactional knowledge. Part of the charge then of the administrator is to develop such relationships. This is a point of significant balance between transparency and confidentiality that needs professional structure.

The administrator needs to also provide condensed information handouts for potential partners in a transaction. Brownfields can be among the most complicated of real estate transactions. The issues can seem overwhelming. Projects can be stalled before they start if the fears seem to be supported by the maze of questions and complexity of process. Governments have developed information handouts with answers for many of the questions that can alarm laypeople. The process can be facilitated by providing these as well as examples of successful resolutions. Some miserable sites have been redeemed. Trust can have a greater chance to develop when the prospective partners can see the road taken by other parties.

The same information is vital to the committee. They need to know that the outcome will have real value. For example, no one would want to participate in a process that appeared to provide an owner with both funding and management of an environmental problem caused by willful or careless mismanagement. Therefore staff needs to do some preliminary research on the regulatory record of the site and the owners before there is a formal assessment. The ultimate beneficiary must be the public. The owner, even if it is a shell corporation, may have an open ended situation become closed to their coincidental benefit, but the asset gain needs to belong to the public. Measurements of that gain could be tax income and jobs as well as environmental remediation. It is important to keep on track to accomplish resolution of the site despite the anger that some members of the public feel about the individuals or company that were responsible for the contamination.

Unfortunately, the research on sites may, on occasion, reveal that the regulators could be seen as having been lax in enforcement. With the ratio of regulatory staff to the number of incidents being what it is, the attention and follow up go to the worst problems, of which there are many with new ones always being added. Therefore, it is useful for staff to discuss with the regulators what protocols and specific areas of concern they have about a site before it is presented to the committee. In this way, everyone can be prepared to answer questions about what can provide a resolution going forward rather than focusing on past events. The trust between the administrator and the regulator is essential if they are to find the points of possible flexibility along with those points of little alternative approach. Getting things done while paying attention to the appropriate resolution of areas of concern relies on knowledge and transparency. Fostering the public good is the charge of all the parties.

When there is agreement on what issues the regulators want addressed and how they want them addressed, the contractors can be asked to provide proposals. Then the committee and administrator can review the proposals. They can perform the comparison of costs and scope of work to determine if the contractor has addressed all the points of the published request. Sometimes the administrator may prefer to have the contractor meet with thc public to explain the project technical approach. Sometimes the administrator wants the contractors or a facilitator to conduct public outreach to gather the reactions of impacted parties.

The Scope of Work is important to clarify as mush as possible before any of the work begins. The expected outcomes are very important to define. The first round of sampling may reveal additional areas of concern that will require additional sampling. Sometimes this recommendation by the technical consultant can appear to be a self serving and make work proposal. That is why it needs to be clear to all parties what is unknown and what is expected to be known after the sampling. The final outcome of the assessment is to determine what the clean up cost is for the proposed reuse of the site. Getting to that answer can require discussion after the completion of different activities to decide on the best next course of action.

For example, the first round of ground water sampling could reveal some contamination. Additional sampling would probably be required to establish ground water flow, the area of the plume, the possible source and all the constituents of the contamination. Sometimes that source can be off site and the release is flowing onto or under the site. That situation may require new access agreements with the owner of the site where the possible source is. This situation can often cause the need for a negotiated legal solution.

The soil sampling may identify areas of concern. The first round of samples may be targeted in the areas of suspected previous potentially polluting activity or in a grid pattern across the site. The next step may be to allocate funds to conduct additional sampling to define the limits of the impacted area in order to quantify the amount of impacted soil. That information is necessary to develop a remedial strategy. It needs to be determined if the soil needs to be removed or if it can be treated in place and the cost of the different alternatives. Being patient while answers develop is important. It can take some time until the lab results are developed followed by the refining of what the contractors suggest doing and finally determining what the regulators want done next. That process may be supported if the administrator is familiar enough with the science and regulations to make suggestions as to how to reach a conclusion. As with any real estate deal, the project must move along. The dialogue costs money and needs to be managed. Often it is important to put a proposed plan in front of the parties for them to react to rather than to wait for them to have the time or to make the time to prepare a plan. They may be able to more quickly respond by editing rather than drafting.

It is also important to be certain that the contractor has continued to follow the guidance of the regulators in the site work activities. Interpretation can cause unnecessary cost if the details are not confirmed.

The administrator must look ahead to what the next steps could be. If there is contamination on site, it will need to be addressed in some fashion. The proposed reuse of the site may help determine how the parts of the remedial plan can be most economically defined. Before creating the Remedial Action Plan, a transparency of the reuse plan could be overlain on the plotted areas of environmental concern. It will be helpful to include ground water flow and possible receptors to reassure people that all the risks have been considered. This will help people visualize where contamination might be isolated and secured on site. Cleanup costs could be significantly reduced.

The work associated with assessing Brownfields will continue as long as manufacturing does. Unfortunately there will always be new sites caused by economic or natural disasters. Technology deployed through more pro-active regulatory operational controls could reduce the spills and dumping and the risks associated with abandoned sites. Elements of experimental Brownfields programs will become institutionalized. A process needs to be in place to assist the municipalities and/or more regional government entities in the resolution of these sites. If that process is open, the public will continue to want their tax dollars used as leverage for clean up and renewal. We are building a process today that will help to resolve yesterday and provide support tomorrow.

Brownfields and urban design: learning from Atlantic Station

R. Dagenhart, N. G. Leigh & J. Skach
College of Architecture, Georgia Institute of Technology, USA

Abstract

Only a decade ago, large brownfield sites were valued only for parks and open space in Europe and for replacement industry or jobs in the U.S. Today, these sites are becoming valued real estate development opportunities for high density, commercial and housing mixed use projects, reflecting the changing form of the contemporary city. If these central city developments are to gain approvals from a myriad of local, regional and national stakeholders, they must be designed to become part of the surrounding city and neighbourhoods instead of continuing their former roles as stand-alone sites. Landscape designs for recent brownfield parks offer some clear directions for understanding urban design for brownfield redevelopment. We relate this landscape experience, such as from Westergasfabriek Culture Park in Amsterdam, to the master planning of Atlantic Station, a high-profile mixed use brownfield project in Atlanta, Georgia. We conclude with three urban design lessons for future brownfield work: subdivide first, design streets as public space, and replace buffers with boundaries to bind the parts of project together to create communities.
Keywords: urban design, landscape urbanism, brownfield redevelopment, Atlantic Station.

1 The new city

For much of the 20th Century, the spatial pattern of American cities was easily understood. Central cores contained downtown commercial districts and dense residential neighbourhoods primarily occupied by the very wealthy and the very poor. Suburbs, containing single-family residences for the middle class, had densities that declined as they extended into the hinterlands. This pattern began changing after WWII, with urban populations decanting to suburban locations

WIT Transactions on Ecology and the Environment, Vol 94, © 2006 WIT Press
www.witpress.com, ISSN 1743-3541 (on-line)
doi:10.2495/BF060181

followed by commerce and industry that created a multi-ring polycentric structure [1]. Central cities were left with the remains of the former city, especially the brownfields that are now the focus of renewed redevelopment. Change has accelerated since the 1980's with many describing this new American city form as a splintered metropolitan network of nodes and links extending across entire metropolitan areas, into exurban territories, and reaching out globally [2].

Density gradients, a simple method of representing population density across urbanized areas, reveal clearly reveal this changed form. Instead of population densities dropping rapidly from urban core to fringe, America's suburbs are becoming denser, while central cities, after a long period of decline, are re-populating [3]. Clusters of apartment communities, office towers, and shopping districts are being developed across entire metropolitan areas at densities that were unheard of even a decade ago. These are mixed with ever-expanding subdivisions of single-family houses in suburban and exurban locations. High density housing, retail and offices, not industry, are now the dominant consumers of available land in central cities. These shifts involve more than just the centre and the far periphery – they also include the older suburban rings which are losing post-war manufacturing to exurban locations, while often gaining new higher density residential developments.

Atlanta, Georgia, widely regarded as the worst example of "sprawl," and one of the best examples of the new form of the 21st Century city, illustrates America's changing urban form. In 1980, downtown Atlanta contained almost 70% of the office space and office jobs in the metropolitan area. By 1990 downtown had less than 30% and is outranked by multiple "edge cities" combining office, retail and high density housing [4, 5]. Today, the downtown share is even less. Manufacturing near Atlanta's central core has vanished and is now disappearing from the older suburban rings. Ford and General Motors recently announced closure of their automobile assembly plants in Atlanta, both of which are located in early Post-War suburbs. Kia, the South Korean automaker, recently announced a new assembly plant in Atlanta's exurban territory, 75 miles from the central core. The City of Atlanta itself, comprising only 10% of the metropolitan population, is now growing after declining for more than 50 years. Atlanta's Beltline, an abandoned railroad ring with adjacent former industrial properties, is now being redeveloped, with estimates of at least 50,000 new residents in the next 20 years [6]. The Beltline alone would increase Atlanta's population by almost 15%, an unimaginable prospect only a decade ago. Atlanta is just one example among dozens of growing cities across the country. The densest American urban areas are now in the West and Southwest – Los Angeles, Phoenix, San Diego – not New York, Boston or Chicago. Even the great 19th and early 20th Century industrial cities, like Pittsburgh, Detroit and Cleveland, are re-forming themselves even though their economies are stagnant and populations still declining [7].

2 Brownfields and the new city

This emerging contemporary city has changed perceptions of brownfield redevelopment. A combination of public political pressure and weak real estate markets directed early brownfield developments in Europe and Canada toward public parks and open space, while in the U.S., the emphasis was focused on industrial reuse and job creation [8, 9]. More recently, the strongest markets and most favoured uses for brownfield redevelopment are housing, retail and office, often in mixed-use formats on large sites. Instead of challenges to be avoided or delayed, brownfield sites are becoming aggressively sought after real estate opportunities. Developers and economic development agencies have gained considerable experience with brownfield developments, and institutional and other barriers to their redevelopment have receded. But it is the changing city form and accompanying real estate markets that are stimulating the search for more sites for new central city and older suburban ring developments. With conventional sites increasingly scarce, brownfield redevelopment are eagerly sought.

Contemporary real estate developments on the urban fringe are understandably designed to stand alone within the ever-expanding node and link form of the new city. They are usually internally focused, surrounded by buffers, and minimally connected to their surroundings. Any positive externalities are closely guarded to minimize market or finance benefits to surrounding properties.

This same design strategy is too often applied to central area real estate developments, especially large brownfield sites, where planning is narrowly focused on site-specific engineering, remediation technologies, immediate marketability of recovered land, and the frequently defensive postures of surrounding communities. The development complexities of large brownfield sites, however, must be balanced in central areas with the necessity for numerous stakeholder partnerships. Multiple local government agencies, public development corporations, adjacent communities, and myriad interest groups are demanding that new developments focus externally, bind themselves to surrounding communities, and provide a dense network of connections to the larger city. Brownfield planning and design must respond to both of these internal and external conditions and to the unpredictable conditions that accompanying them during the planning, design and development process.

3 Brownfields and landscape design

Design has played an important role in brownfield development since the 1980s, but the emphasis has been primarily on parks and open space [10]. When one thinks of design and brownfields, large park proposals immediately come to mind. The well-known park projects in Paris – Parc de la Villette and Parc de Citroen – were both developed on brownfield sites. Others include Landscape Park Duisburg-Nord in the Nord/Ruhr District, Westergasfabriek Culture Park in Amsterdam, the Sydney Olympic Park, and a long list of others, most of them in

Europe and Canada [11]. The Fresh Kills Landfill on Staten Island is one of the few U.S. examples, and it has only recently begun [12].

Design of reclaimed land for parks has played a key role spawning a design approach called Landscape Urbanism [13]. Instead of setting out a visual image of a proposed park or open space, the focus is instead on process. Derived from ecological concepts, Landscape Urbanism focuses on defining a precise organization of territory as a framework to relate the expected changes – in infrastructure design, remediation processes, park programs, financing sources, etc. – during the entire process of development. This organizing framework consists of three major parts: defining boundary conditions, connecting the site and its surroundings, and internally organizing the site itself, especially circulation channels and major site features.

Amsterdam's Westergasfabriek Culture Park illustrates this Landscape Urbanism design strategy [14]. The park framework design was negotiated with the adjacent neighbourhood, the municipal planning bureau, the developer, and other regulatory agencies and interest groups. It included four key parts: (1) definition of each of the park's boundaries; (2) locations of connections to the adjacent Westerpark, the adjacent street grid and the polder on the opposite side; (3) the location of an central circulation axis linking the remaining buildings together; and (4) a large green space to be used as a multi-use event field. This agreed upon framework – a simple diagram – allowed detailed planning and development of the park to include typical unforeseen changes, such as altered remediation requirements and technologies, changing uses and programs, and altered infrastructure financing and design. In short, designing the Westergasfabriek Culture Park, like the other brownfield parks mentioned above, was like designing part of a city with participation of its many stakeholders, clients, technicians, regulators and finance agencies. The design framework was both specific and flexible. It allowed changes to occur over a ten-year development period and will continue to do so as the park is fully occupied and matures in the future.

4 Brownfields and urban design

Urban design can be approached in many different ways. Too often, it is simply a land use plan – a diagram showing the locations of different project zones and the streets that link them together. At other times, the focus is a visual theme, branding a development with a particular architectural style. And still others, it may focus on building arrangements creating a picturesque scene, a small-town ensemble, a campus-like setting and so on.

When brownfield redevelopment aims for housing and commercial uses, rather than parks or industry, urban design is best understood as a corollary of Landscape Urbanism. Just as Landscape Urbanism adopts an ecological metaphor for designing parks and other large public spaces, urban design adopts the metaphor of traditional city development processes, especially the American process of subdividing land into lots, blocks and streets to make a subdivision plat or city plan. Eighteenth Century Savannah, 19th Century Manhattan, and early 20th Century suburban subdivisions across America follow that process. The image or style of their streets and public spaces change, buildings are built and re-built, and uses change, too, but within the established subdivision framework. It is a very simple idea, but too often ignored in the face of complex political, market, finance and technological pressures. Urban design, then, can best be understood as a two-part process.

First, urban design is about organizing territory – creating a subdivision plat of lots, blocks and streets – to establish a framework for development. This subdivision process, like the landscape urbanism frameworks, consists of three major parts: defining each boundary, connecting the site to the surrounding city, and organizing the site itself into lots, blocks and streets. This subdivision plat must be designed collaboratively with the full complement of stakeholders.

The plat is then the template for multiple design, finance, construction and leasing/sales decisions. This includes infrastructure design, development approvals, development programs, changing development partnerships, and remediation technologies. It also includes long-term changes as uses change, as buildings are demolished and rebuilt, and as streets and public spaces are redesigned to fit new circumstances. The subdivision plat serves as both a compressed version of traditional city building processes in the short term and a parallel to traditional cities in the long term.

5 Brownfields and urban context

Brownfield sites are both small and large – the gasoline station or dry cleaners as well as the large former industrial sites. However, more important than size is their urban context. Some brownfield sites are small parcels located within an existing urban framework of lots, blocks and streets. Developers of these urban sites now anticipate some level of remediation as part of a normal development process. Often these small sites are part of a former industrial or warehouse zones that have created brownfield districts. In these situations, the redevelopment process is clear – the prior lot, street and block urban subdivision plat exists and provides the framework for incremental redevelopment, including incremental remediation, infrastructure improvement, marketing, and so on. The well-known South CBD District in Chattanooga, Tennessee is a brownfield redevelopment district of this type [15].

Many of the most sought after brownfield sites, given the changing structure of the new city, are the larger sites once occupied by large industries. These sites were usually located for railroad access, with only tenuous connections to the surrounding lot, block and street framework, although they may be adjacent to residential neighbourhoods, Redevelopment of these large sites requires large-scale land acquisition, extensive remediation, major front-end infrastructure investments, and complex dealings with local governments, regulatory agencies and neighbouring residents and property owners. If these projects are simply returned to industrial uses with continued rail access, then development can usually move ahead quickly. However, in the newly developing city, where housing, retail and office are preferred, these railroad territories must be made to be "urban." Atlantic Station in Atlanta, Georgia, developed on the site of a former steel mill, is a well-known and high-profile brownfield development of this type [16].

6 Atlantic Station – urban design and brownfield development

Atlantic Station, a 138 acre (56 ha.) site, located at the intersection of Interstates 75 and 85 on the north edge of central Atlanta, was formerly occupied by the Atlantic Steel Mill. With an investment of more than $2 billion, Atlantic Station is cited as the largest brownfield redevelopment project in the U.S. A joint venture of Jacoby Development of Atlanta and AIG Global Real Estate, the final build out of the development will include 6 million square feet (550,000 square meters) of office space, 10,000 new dwelling units, 1.5 million square feet (140,000 square meters) of retail space and 1000 hotel rooms. Begun in 1997 and opened in the fall of 2005, Atlantic Station is praised as a model of the U.S. Smart Growth movement, receiving the Phoenix Award from the U.S. Environmental Protection Agency (EPA) as the nation's best brownfield redevelopment. Atlantic Station has won praise from a combination of unlikely sources, including the Urban Land Institute and The Sierra Club. As real estate investment, Atlantic Station is highly successful to date, which is a substantial accomplishment for such a large scale and complex brownfield development.

Much of the praise for Atlantic Station is for its master planning, which began in 1997 and continued for six years through negotiating and gaining approvals from stakeholders. Although laudable for its moving a complex project ahead, the master planning process at Atlantic Station offers important and cautionary urban design lessons that to apply to other large-scale brownfield redevelopments. The planning process involved Jacoby Development's original plan, two subsequent revisions, and a final one that was approved for construction with minor changes [17].

The First Jacoby Plan, fig. 1, was a conventional, suburban-oriented land use based master plan. Three areas were designated for large single use developments by separate developers. Each of these projects – a horizontal shopping mall, apartment complexes and an office park were automobile oriented, internally focused, surrounded by surface parking lots, and separated from each other and the surroundings by green space buffers. The three areas were linked together by a new 17th Street connecting Northside Drive on the west and on the east across a new bridge spanning I-75/85.

The First Revised Jacoby Plan, fig. 2, followed the developer's realization that rezoning, site remediation and bridge construction over the Interstate Highway required collaboration with numerous stakeholders. Satisfying these stakeholders led to adoption of Smart Growth principles to redirect the project to higher densities, mixed uses and increased pedestrian and transit accessibility. This plan formed the basis of City of Atlanta zoning approvals, approval of tax allocation district to finance site remediation, and a proposal to EPA for Project XL status that gives the developer flexibility in meeting EPA requirements in exchange for adopting innovative ways to protect environmental and public health. The proposal to EPA was critical, not only to speed environmental approvals but also to gain permission and allow U.S. Department of Transportation (DOT) funding for the 17th Street Bridge. At the time, Atlanta was under an EPA moratorium for highway construction funding due to non-compliance with the Federal Clean Air Act. Although the appearance of the Jacoby plan changed, the initial land use based diagram remained. Modest mixed use developments were included, and minimal street connections were made to the Home Park neighbourhood to the south. The major change was placement of an 8000 car parking deck under the entire retail district. Although claimed as a logical step to cap part of the contaminated soil on the site, it appears that the parking deck was suggested first and foremost as a plinth for a traditional town-like retail district on top.

The Duany Plater-Zyberk (DPZ) Plan, fig. 3, resulted from EPA requiring Jacoby to bring in additional consultants who had more experience with Smart Growth projects than Jacoby's architects. The DPZ Plan discarded the land-use oriented plan and created a lot, block and street subdivision, with fine-grained mixed use across the site. This subdivision structure continued the Home Park neighbourhood street and block format into the Atlantic Station site. The DPZ plan also minimized the underground parking decks by incorporating parking into the centre of blocks and locating underground decks only where necessary for capping contaminated soil. All stakeholders applauded the DPZ Plan, especially for its reduced automobile dependence, connections to the surrounding neighbourhoods, and the framework extending across the whole site instead of three separate parts.

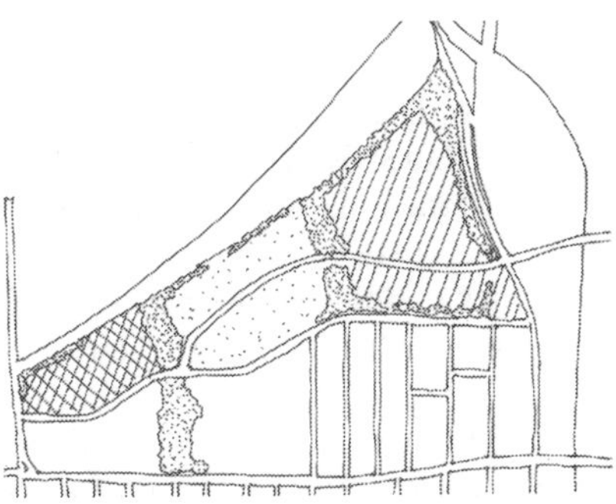

Figure 1: First Jacoby Plan.

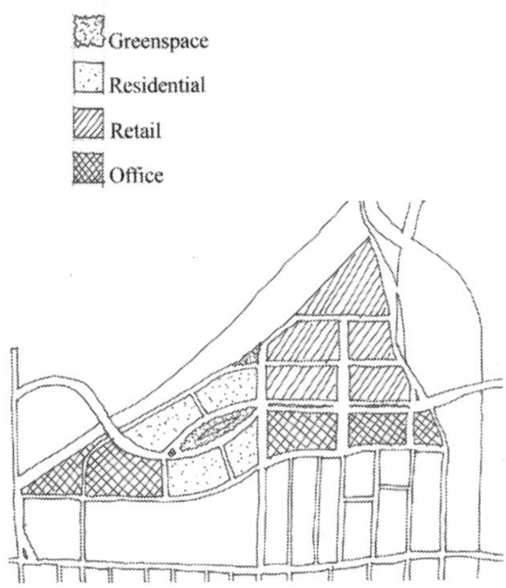

Figure 2: Revised Jacoby Plan.

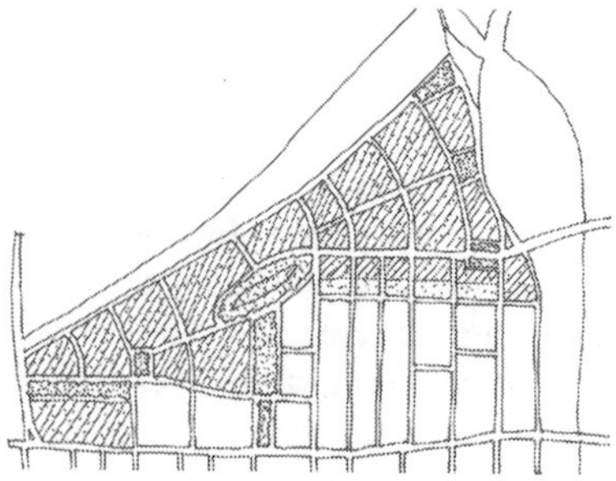

Figure 3: Duany Plater-Zyberk Plan.

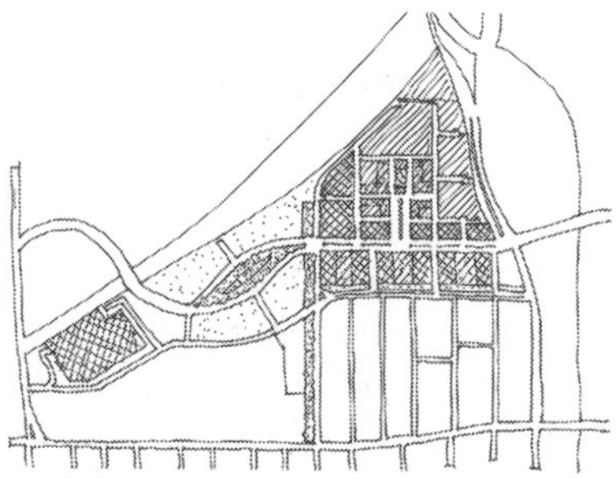

Figure 4: Final Jacoby Plan.

The Second Revised Jacoby Plan, fig. 4, was the response to DPZ's recommendations. It claimed to incorporate many of the recommendations, but did so modestly. The massive two-level parking deck under the retail district remained. Smaller dimensioned blocks were introduced to produce a network of local streets, although their connections to the surroundings were minimal. A more fine grained mix of uses was introduced on a few block frontages. Home Park streets were shown, but only with minor connections to the site. Importantly, the Jacoby's original land use based organizing diagram remained with the new 17th Street and bridge remaining the dominant feature of the development.

Since its opening, Atlantic Station has been praised for its real estate development prowess, but criticized for its urban design shortcomings. These criticisms are focused primarily on three aspects of the project and linked directly to the master planning process.

First, the internal subdivision is based on three primary land-use zones, remaining from the First Jacoby Plan, fig. 1. Even the buffers remain between the three zones, such that it is difficult to walk rather than drive between the apartment complexes in the central area to the commercial district of the IKEA store that replaced the office park on the east. The small block structure in the commercial district is designed primarily for its image as a urban place, not its primary structure that occurred in traditional cities. The streets lead nowhere; access to the massive parking structure below the whole district, are not part of the gridiron of streets. The blocks are also dimensioned, according to building types, essentially sized for one building per block. This is the opposite of the DPZ Plan and traditions of urban design where block dimensions are not detailed for specific uses and, instead, can easily accommodate building and land use changes over time.

Second, the internal connections are poorly designed as part of an urban, mixed-use environment. The bridges and 17th Street were designed according to DOT standards, not according to urban design or context sensitive design standards. Originally four traffic lanes with a wide median, 17th Street was expanded with dedicated bus lanes, dedicated bike lanes, and dedicated left turn lanes that practically eliminate pedestrian waiting space at crossings. It is so oriented to automobile flow that trees are set back eight feet from the curb to eliminate hazards to automobiles, while on-street parking eliminated, putting pedestrians next to fast moving traffic. Sixteenth Street at the boundary of Home Park is designed similarly. Instead of being a central feature of Atlantic Station, it is a barrier slicing it the development in half.

Third, Atlantic station does not bind to its surroundings. This is especially true on the south boundary where it joins the Home Park neighbourhood, a historic neighbourhood of modest houses where steelworkers once lived. The continuity of streets, proposed in the DPA plan, are disjoined. Although presumably protecting the neighbourhood from cut-through traffic, residents find it almost impossible to walk to the supermarket, restaurants, or offices in Atlantic Station. Thus, the Smart Growth features of Atlantic Station that qualified it for significant public subsidy only benefit newcomers and not the historic residential community adjacent to the former steel mill. And from a design perspective, Atlantic Station presents barriers on each side. Buildings turn their backs to the Interstate Highway and to the railroad and Loring Heights to the north. The auto-oriented IKEA store that replaced the office park on the west side eliminated any connection other than for private automobiles accessing loading zones or the underground parking deck. The 16th Street boundary with Home Park presents a long row of townhouses, effectively making a visual barrier between the neighbourhood and Atlantic Station.

7 Conclusions

Definitive conclusions regarding the ultimate success of Atlantic Station are premature. It will take 10 years to complete, according to current forecasts, and many more years to mature and begin the processes of changing tenants, new uses, new buildings and new designs for streets and open spaces. However, there are three lessons that are useful for other large brownfield projects of the scale and complexity of Atlantic Station.

First, the pedestrian unfriendliness of Atlantic Station reinforces the wisdom of traditional urban design practice where the large scale development site is subdivided first. In this practice, the structure of lots, blocks and streets is not determined by land uses but is, instead, governed by knowledge of cities and districts that have proven themselves through time.

Second, streets should be designed as the most important public spaces because they are, indeed, precisely that. Although streets must accommodate automobiles, pedestrians, public transit, and bicycles, their design should always have an objective of creating public space, as structured within the subdivision framework. When designing streets, it is always valuable to measure and observe great streets that work from their history.

The third lesson is actually a warning, and one that is especially warranted for the complex undertaking of brownfields. It is important to understand that the first diagram developed for a project will be resistant to change, because even if preliminary, it has already had commitments built into it. Thus, from the beginning, the design process must be collaborative and completed without haste; it will change only with colossal effort and will probably, like Atlantic

Station, appear strongly at the end of the planning process. In the maturing of the practice of large scale brownfield redevelopment, developers have come to expect significant public assistance such as that received by Atlantic Station. But without a commitment to the collaborative planning and design process from the beginning, the extraordinary public assistance for a private development project may not be justified, and the potential to be a model urban redevelopment project that has positive externalities on the broader community will not be realized.

References

[1] Lee, S. and Leigh, N.G., The Role of Inner Ring Suburbs in Metropolitan Smart Growth Strategies, *Journal of Planning Literature*, 19 (3), pp. 33-46.

[2] Graham, S. and Marvin, S., *Splintered Urbanism*, Routledge: London, 2001.

[3] Bruegmann, R.., *Sprawl: A Compact History*, University of Chicago Press: Chicago, 2005.

[4] Garreau, J., *Edge City*, Doubleday: New York, 1991.

[5] Leigh , N.G., People versus Places: Telecommunications and Flexibility Requirements of the CBD, Cities in the Telecommunications Age: The Fracturing of Geographies, Routledge: London, 2000.

[6] Beltline Partnership, http://www.beltline.org/what/economicDevelopment.shtml#redevelopment

[7] Bruegmann, R., *Sprawl: A Compact History*, University of Chicago Press: Chicago, 2005.

[8] De Sousa, C. A. Turning Brownfields into Green Space in the City of Toronto, *Landscape and Urban Planning* 62, pp. 181-198, 2003.

[9] United States Conference of Mayors, Recycling America's Land: A National Report on Brownfields Development., United States Conference of Mayors, Vol. 3, 2000.

[10] Kirkwood, N., Integrating Technology and Design in Reclaimed Landscapes. *Manufactured Sites*, New York: Spon Press: New York, pp. 3-11, 2001.

[11] Kirkwood, N., *Manufactured Sites*, Spon Press: New York, 2001.

[12] Field Operations, Fresh Kills Landfill to Landscape, *Praxis* 4, pp. 18-27, 2005.

[13] Waldheim, C., Landscape Urbanism: A Genealogy, *Praxis* 4, pp. 10-17, 2005.

[14] Koekebakker, O., *Westergasfabriek Culture Park*, NAI: Rotterdam, 2003

[15] South CBD Plan, Chattanooga, Tennessee, www.rivercitycompany.com/pdfs/media/southside_redevelopment.pdf

[16] Atlantic Station, www.AtlanticStation.com

[17] Transportation and Environmental Analysis of Atlantic Steel Development, www.epa.gov/dced/topics/atlantic_steel.htm

Section 7
Lessons from the field

Defence heritage moves on to civilian futures

C. Clark
The Wessex Institute of Technology, UK

Abstract

Significant historic sites associated with national defence are on the closure list in many countries, as armed forces are reduced and regrouped. This paper draws on recent experience of their conversion to civilian uses in the US and Europe.

Government and independent research devoted to this specialised area has appeared in England, France, Germany and the US, although the topic is still rarely explored. In the 1990s, English Heritage produced thematic studies of dockyards, barracks and subsequently airfields to identify important examples of building types worthy of preservation. The UK Ministry of Defence produces annual Stewardship reports on the defence estate. In 2002 the French Ministry of Defence celebrated its heritage in a lavishly illustrated book: *Patrimoine militaire* (Editions Scala, Paris). The Bonn International Center for Conversion (BICC) does pioneering work on decontamination and the economic effects of closure upon defence dominated communities. In the 1988–1995 rounds of the US Base Realignment and Closure process used by the Department of Defense were used to review and restructure its military installations, and the US House Armed Services Subcommittee on Readiness have held hearings on the management of historic military properties and installations.

Countries differ considerably in the regulations which govern the transitional process to new uses. How defence sites are disposed of – by free transfer, development by the local authority, public private partnership, or sale to the highest bidder – has a strong influence on the power of different parties to influence decision-making and hence on the outcomes and new land uses.

This rapidly moving and widespread process is still not much studied. There is little interchange between countries undergoing similar experiences, the importance of historic defence sites to their local communities is both cultural and social. More needs to be done to share good practice in this most important of transitions.

WIT Transactions on Ecology and the Environment, Vol 94, © 2006 WIT Press
www.witpress.com, ISSN 1743-3541 (on-line)
doi:10.2495/BF060191

1 What is being disposed of?

Significant historic sites associated with national defence are on the closure list in many countries, as armed forces are reduced and regrouped. In the 1988-1995 rounds of the US Base Realignment and Closure process used by the Department of Defense to review and restructure its military installations, 97 major bases were closed, and there are plans to close 25 more and 'reconfigure' 26 others in the current round. Included are National Historic Landmarks: Fort Monroe in Virginia, Fort McPherson in Georgia, Vancouver Barracks in Washington and Fort Haye in Ohio. In the UK Hounslow Barracks, Haslar Hospital and HMS Daedalus are amongst the historic properties to be sold, but also for sale are firing ranges – brownfield sites par excellence [1].

2 Research into the history, condition and future of defence sites

Understanding the history of defence heritage sites and maintaining them in good condition are essential first steps in their move towards civilian futures. Research in England identifies and celebrates the range of defence structures identified as heritage. The British Ministry of Defence is responsible for the largest number of historic environment assets on the UK government estate, with approximately 650 listed buildings and over 1,000 scheduled monuments [2]. In 1998 English Heritage produced thematic studies of dockyards and barracks to identify important examples of building types worthy of preservation [3]. In 2004-5 they conducted a study of the development of military aviation from its inception and up to 1945. The aim was to ensure that airfield sites are adequately recorded and protected, and to document their historical context for management and educational purposes. This study culminated in December 2005, with the listing of 255 buildings, 86 of which are on the defence estate on 19 MOD sites. The extensive restoration of the UK Ministry of Defence (MOD) Main Building in Whitehall London, constructed over the period 1938 to 1959 [4] is featured in the Department of Culture, Media and Sport's forthcoming Biennial Report. However, the Stewardship Report on the Defence Estate 2005 refers in its section: 'Taking account of the wider interests of society with particular reference to sustainability and the environment'- to defence historic buildings at risk The MOD has committed itself to reducing their number. In 2001 there were 43 at risk: 6.1% of the total; this has now been reduced to 32. However, 33% of the 9000 Scheduled Ancient Monuments on MOD land were in fair condition, while 21% in poor or unknown condition [2]. MOD advisors on the Historic Environment have expertise in both historic buildings and archaeology.

A major enhancement in civilian input into UK planning of the future of defence sites is likely to follow abolition of the Crown Exemption, which exempts government departments from civil planning law. In 2005 the Office of the Deputy Prime Minister drew up the timetable for its removal under the 2004 Planning and Compulsory Purchase Act in 2006. But whilst this removal applies to town and country planning and listed buildings, it does not apply to scheduled monuments, which is an omission of the 1979 Ancient Monuments and Archaeological Areas Act. However, this will be rectified as part of the DCMS Designation Review which is streamlining the designation system, but in England only. Scotland and Wales have separate heritage organisations [5]

In March 2006 the US House Armed Services Subcommittee on Readiness held a hearing on Historic Military Properties and Installations to hear testimony on the management of historic military properties and installations by the Armed Services. Phil Grone, the deputy undersecretary of defense for the environment and installations outlined the Department of Defense's (DoD) program to manage historic properties. DoD has management responsibility for 75 National Historic Landmarks – as well as nearly 600 historic entries listed on the National Register of Historic Places – which encompass more than 19,000 individual historic properties located on over 200 military installations. The DoD has a broad strategic framework for managing the agency's historic properties that includes early planning and the use of Integrated Cultural Resource Management Plans (ICRMP) which not only identify the inventory of assets (historic buildings and structures, traditional cultural places, archeological sites), but also include the actions needed to manage and maintain those assets. Interestingly, in view of the common prejudice that older properties cost more to maintain than modern buildings, Grone also noted that DoD studies have reinforced the conclusion that "on a per square foot basis, the costs of maintaining Historic Properties are frequently comparable to the costs of maintaining non-Historic Properties." He cited GSA statistics which "found that operating costs for the historic buildings in the study averaged 10 percent less per rentable square foot than non-historic properties, with utility costs 27 percent less", due to superior construction [6].

The Bonn International Center for Conversion (BICC) does pioneering work on decontamination and the economic effects of closure upon defence dominated communities in many countries including Germany. As the bibliography in my study of European dockyards demonstrates [7], however, literature on post-defence reconstruction, its economic, social, spatial and physical impacts, and the implications for the historic defence estate is rare, though more common in the US. Rachel Woodward in *Military Base Conversion and Rural Development* [8] says that while the effect of conversion on the defence industry has dominated conversion research, military base conversion is the Cinderella of conversion studies. An example is *Restructuring The Global Military Sector Volume II: The End of Military Fordism* [9] which is mainly focussed on defence conversion, although there are regional studies on Ukraine and the Philippines. Because of the visibility of defence industrial restructuring in economic and policy terms, and the position of the defence industry as connector between security and the economy [10], the conversion of military properties and lands such as airfields, depots, barracks, dockyards, training centres has received far less attention. As Broska [11] notes, this absence in social scientific research has left significant gaps in our knowledge about the effects of base closures and the prospects for civilian reuse in many parts of the world. Britain is a case in point. There are two reasons for this lack of critical appraisal. Although base conversion may be intermittently highly visible in land use and development policy debates, for example, following publication of the UK *Strategic Defence Review* [12] and the *Comprehensive Spending Review* [13] administrative and academic interest is not sustained because of the long timescales involved [14]. In contrast is the sustained work of the Bonn International Center for Conversion, which since 1994 has been actively tracking all aspects of the conversion processes in central and eastern European countries; for example military expenditure, surplus weapons, demobilisation, conversion of military research and development, conversion of defence industry and base conversion. It "facilitates the processes whereby people, skills, technology and economic resources are shifted away from the defense sector and applied to alternative civilian uses" [15]. Their publications include their *Conversion Surveys of Global Disarmament, Demilitarization and Demobilization* of 1996, 1997, 2002 and 2003 [16], papers on Base Closures in Hungary and Poland, and an extensive study commissioned by the German Federal Ministry of Environment [17]. Henri Myrttinen's study of central and eastern Europe [18] and the BICC Brief 18 on *Conversion in South Africa in the 1990s* [19] offer detailed analysis of experience.

3 The effects of disposal systems on outcomes

Countries differ considerably in the regulations which govern the transitional process to new uses, with consequent effects on decision-making and on the social, environmental and economic outcomes. The emphasis in the United Kingdom is on achieving a high financial return to the Treasury, a stipulation which has considerable effects on the subsequent land uses. The Ministry of Defence estates agency has a requirement by the Treasury to achieve maximum financial receipts from disposal of defence property. For the current year this is £151 million [5].

High end housing and leisure shopping bring developers who have paid a high price the highest return. The sale price may be slightly reduced for historic sites according to guidelines for government departments by the Department of Culture, Media and Sport. In contrast, the US Department of Defence (DOD) has a federal preservation officer whose task is to convey to local communities whose economy may be blighted by defence cuts how historic properties can be an engine for redevelopment. Objective reviews of each sites and public hearings were held before the 2005 recommendations on closures were made to the President. The DOD Office of Economic Adjustment helps local communities plan new uses for surplus properties.

4 Public/private partnerships

In Britain, where there is a government target for 60% reuse of brown land, a new development is the reliance on public/private partnerships to bring unused historic defence areas back into use. Perhaps the MOD's desire to retain receipts from property disposals rather than see them revert to the Treasury underlay the new strategy for Portsmouth dockyard in southern England. Portsmouth Naval Base is still the berthing port for many naval sips: carriers, Type 23 frigates, Type 42 destroyers, mine counter measure vessels and patrol ships, with facilities to berth visiting international ships. In view of a continuing need for defence spending, Project Prime (Portsmouth Regeneration in the Maritime Estate) was launched in 2004/5 after three years' gestation. It represented a new model of naval base renewal for the British Ministry of Defence. Although it was cancelled in July 2005, it is described in detail here because it illustrates the complexity and sheer difficulty of achieving sustainable naval waterfront renewal, especially when the navy is still operational next door. Project Prime's aim was:

> "to generate income [from the private sector] to invest in a stream-lined Modern Naval base, directing resources in support of the [national defence] front line, by the creation of residential, leisure and commercial opportunities, providing employment opportunities within the local area, and preserving and enhancing buildings of National Historic Importance" (my insertions) [20]

In October 2001 Defence Estates, the MOD's property agency, was commissioned to deliver a master plan for the naval base. This document identified an area of approximately 30 acres in the south west corner of the base capable of release for commercial exploitation, whilst maintaining the operational effectiveness of the dockyard of 333 acres. This estate lies within a conservation area containing listed buildings, ancient monuments and other historic structures. A small area of what was under consideration has been leased for 99 years since 1985 to the Portsmouth Naval Base Property Trust, but the dowry which accompanied the setting up of the Trust was nowhere near enough to bring all the old buildings back into repair or to fully convert them to new uses. Some of the empty space in the masthouse, stores-houses and boathouses has gradually been refurbished and reused for the Royal Naval Museum, the Mary Rose Trust, dockyard trades display and the interactive Action Stations visitor attraction, as well as a café and art gallery, but fundraising for each project was time consuming and expensive, and upper floors remain unused.

The aim of Project Prime for the MOD was to use capital raised through the release of a larger area to civilian capital investment to re-invest the proceeds in the MOD estate, re-providing facilities lost within the hived off area with modern equivalents in the naval enclave. It would also have relieved the Naval Base Commander's budget of the high maintenance cost associated with the historic estate [21]. It was hoped 'to save upwards of £25 million. Naval Base Commander Cdre. Amjad Hussain said that 'buildings such as the Old Naval Academy, Anchor and Stony Lanes, Short Row and the Block Mills had served the Navy well, but no longer met the needs of the 21st Century fleet' [22]. Funds from Project Prime would not have disappeared into the 'black hole' of the Treasury, but have been reinvested in modern facilities such as offices in other areas of the naval base [23] and restoration of the internationally important Block Mills, currently at the top of English Heritage Buildings at Risk list.

A single Public/Private Partner was being sought to provide a "holistic" solution for the entire Project PRIME development area by an integrated project team from Defence Estates, the Naval Base Commander organisation, supported by Partnership UK, who are co-sponsoring this project. Partnership UK is a Public/Private Partnership stemming from the HM Treasury Private Finance Initiative Taskforce that develops the private finance agenda and 'Wider Markets' initiatives on behalf of government – to achieve enhanced value by going to the market [21].

The assets of the Property Trust and the MOD would have been taken on by a PPP (49% Public/51% Private Partnership) to provide new facilities for the naval base and open up 60 listed buildings into the public area, offering them to a developer. Most are Grade II, but Block Mills is a Scheduled Ancient Monument and Grade I, and, shockingly, as already said, it is also No. I building of the English Heritage list of MOD Buildings at Risk. Very large financial resources are needed for restoration and repair, and for conversion to new uses. Instead of the tight security of a naval base, there would have been open access to the public – to buildings locked away for 300 years. The intention was to appoint one developer to take on all the buildings – the majority of which will be profitable conversions to residential use – for example the Naval Academy and Ropery - which might provide cross subsidy for others, such as Block Mills.

Civilian expertise was crucial. The MOD had advice on planning, as well as drawing on commercial and technical advice on the economics and viability of Project Prime from King Sturge (commercial agents) and ENTEC (multi-disciplinary practice), The project was led by Defence Estates, English Heritage, the Naval Base Property Trust and Portsmouth City Council planning, economic development, engineering and conservation officers [23]. There are currently 500,000 tourist visits each year and 375,000 people paying to visit the preserved ships, museums and Action Stations, but the Trust would like to move away from reliance on tourism into a more diversified economy, open in the evenings instead of the site being closed in at night within naval security.

The MOD also draws advice from Defence Estates' Sustainability and Conservation group based in Warminster, which is responsible for conservation of buildings, wildlife, sustainability, energy efficiency, and green eco-building regulations. Up until recently the MOD's tactic was to sell redundant property, but according to Defence Estates, this doesn't necessarily bring in the best return. Although the MOD had not done so before, Public/Private Partnerships have been used by other government departments: for example British Waterways' Isis project, where the whole of the waterside estate including lock keeper's cottages and land was leased away to a developer. Another example is One North East where Newcastle City Council and Gateshead have passed over their properties [21]. Project Prime and Defence Estates MODE London, which used a similar scheme to reduce MOD's footprint in London from 13 sites to three or four, were intended as showcase projects for this type of procedure for the MOD. If it had been successful, the Project Prime model would have been rolled out elsewhere as a mechanism for other surplus MOD land.

The Project Prime developer was to have been engaged on a contractual basis. The MOD would have retained the freehold; the Property Trust would have surrendered its current lease of the Heritage Area and taken on a 125 year lease for the larger Project Prime area, carrying on with its guardianship of the site. It would have been responsible for ongoing repairs and maintenance of buildings and continue its role as their interpreter for the benefit of the public. The developer would have been granted a lease of one day less than 125 years, and paid ground rent to the Trust; they would then have realised (or 'stripped out') the value of the converted buildings. Owners of residential development would have had long leases, which might have been reviewed in the light of new legislation on tenants' right to buy under the new title of commonhold, which brings the UK into line with Europe, Canada, Australia and the US [24]. In view of the several layers of heritage protection on existing structures and lack of cleared land, there would have been only limited facilities for rebuilding and new construction. In March 2005 a design brief – or Memorandum of Information – was discussed with developers [25]. 67 companies expressed interest in response to the announcement in the EU official journal in September 2004: contracts have to be open to all member states. All the companies were British. Berkeley Homes, which had redeveloped Gunwharf Portsmouth, Royal Clarence Yard Gosport, Woolwich Arsenal and other military sites, and Crest Nicholson who had permission for a very large housing scheme on the adjoining Heritage Car Park were not in the final six. Vosper Thorneycroft Shipbuilding who built and operate the vast ship assembly plant built inside the dockyard which had transferred from Woolston Southampton was part of a consortium, reflecting the huge scale of the project. The front runners were utility company Amec, and housebuilders Persimmon who were also in a consortium. The Naval Base Property Trust was part of the two-week sifting process. The developers were shortlisted via a questionnaire and evaluation in a formal procurement process – on their experience, track record, historic conversion, MOD contracts, health and safety, sustainability and long term issues, awarding a graded score [23]. The next stage was to invite the six to submit further details to the Industry Day in June, but the project was cancelled in July 2005. Details of potential uses and their physical implications for the historic buildings are discussed in the chapter on heritage.

It was true to say that Project Prime did not enjoy unmixed support; senior naval officers who dominate the Navy Board and deplored recent defence cuts would have lost their valued social facilities at the Naval Academy; and Admiralty House, which is occupied by the Second Sea Lord. Both were included in the area on offer. In July 2005 after the two suicide bombers' attacks on London and changes at the top of the navy, as already said, the £50 million project was cancelled. The Navy Board concluded it was too risky to go ahead with large scale development by Project Prime, considering that Portsmouth was likely to be the place where the new breed of supercarriers were to be berthed. The proposal to put hundreds of flats and offices near the planned berths was deemed too much of a security risk. "We have two conflicting plans…There's no choice between the two because the first [the berth] is important for the future of the navy and the development of this region's economy as a whole" [26]. As there are existing tall blocks of flats nearby and a very large new complex was granted permission just over the dockyard wall at the same time, as well as the new flats arising on the Gosport shore across the harbour, this statement seems a touch disingenuous.

Another factor was that Project Prime stacked the upfront costs enormously high onto the private developer. The Ministry of Defence wanted all their new offices to be constructed up front, to condense the operational area within more secure and easily defensible boundaries – before the development started and there was any financial return from residential redevelopment of the historic buildings, the likely highest earner. This put developers off – they would have had to take too high a risk. The Naval Base Property Trust advised Defence Estates that Project Prime would have a struggle to be viable because the MOD wanted so much office accommodation before the developers got their money back. There might be an ultimate end profit, but return on capital employed would have been deferred too long [23].

Since the cancellation of Prime, the boundaries of the residential area have been redrawn 300m from the jetties. Defence Estates considered how to use the historic buildings on a one by one basis. To do nothing is not an option, because the historic buildings are unusable in their present state. Instead of a big bang, there's a return to an evolutionary process. After the cancellation, Option 1 was to identify other defence departments wishing to occupy the buildings to be released, but they don't lend themselves to use, for example as offices. Option 2 was to apply to the Chief of Fleet Support – to ask that since buildings occupied by residents in such close proximity to ships raised security issues, what uses are acceptable: offices, bars, restaurants, storage, friendly MOD tenants, Vospers, Fleet Support Limited, a school? Security is changing all the time; the naval base cannot be freed from perceived threat.

A critical problem is that sustaining the complex infrastructure of historic naval sites for civilian futures requires careful long term provision once this is no longer the responsibility of the defence budget. Maintenance of dock walls, culverts, basins, caissons, cranes and other infrastructure would have been separately funded, via a sinking fund which service charges to the new occupiers would not cover. A precedent for this is the Chatham Maritime Trust which was established to look after and collect charges for maintenance – on the advice of Price Waterhouse who analysed commuted maintenance charges which were not included in service charges. This advice was commissioned by the South East England Development Agency and English Partnerships which took on part of Chatham dockyard [23].

But despite the cancellation of Project Prime, change to a new, mixed government/civilian future has already taken place in Portsmouth dockyard, which was ground-breaking in itself. Large areas had already been leased to defence contractors VT Shipbuilding and BAE who jointly own Fleet Support Limited. These might be regarded as friendly users, but they work on civilian as well as naval contracts to ensure continuity of work. VT built a huge new ship assembly shed over No. 13 Dock and BAE converted two floors of the listed Iron and Brass foundry.

Other public/private partnerships may not result in maritime reuse: in Karljohansvern in Norway where there are about 90 buildings older than 1940 considered as historic buildings, most of the buildings directly connected to shipbuilding are in private use, such as smaller houses used as offices and residences. A private company used the dock and several buildings from the late 1960s to the 1990s. Parts of the fortifications were destroyed when the shipyard extended its activity early in the 1970s [27].

5 Conclusion

The lesson from Portsmouth and from other examples such as Charlestown Boston is that the private sector cannot do everything. Some public investment is required for the start-up phase until other sources of income emerge. If the public sector cannot justify its grants for waterfront redevelopment projects as profitable real-estate ventures, then political and urban design criteria should also be included in their assessment [28].

This rapidly moving and widespread process is still under-researched, and there is little interchange between countries undergoing similar experiences, except within Europe, where the EU has financed exchanges between affected local governments. Military geographies are everywhere; every corner of every place in every land in every part of the world of ours is touched, shaped, viewed and represented in some way by military forces and military activities [29]. When the military leave, the importance of historic defence sites to localities is cultural and social as well as economic. More needs to be done to share good practice in this most important of transitions for former defence heritage sites.

References

[1] Wintour Patrick and Elliott Larry 2006 'Brown taunts the Tories: we invest, you cut taxes Education and environment made priorities but no more for health' *the Guardian* 23 March 2006 page I

[2] Ministry of Defence 2006 *The Stewardship Report on the Defence Estate* Ministry of Defence London

[3] English Heritage 1998 *Barracks Thematic List Review Summary Report and Recommendations* English Heritage London English Heritage 1988 *Royal Naval Dockyards Thematic List Review Summary Report and Recommendations* English Heritage London

[4] CTS Media Bath n.d. *Your Guide to Main Building*

[5] Hills-Spedding Guy 2005 Defence Estates Email on Crown Exemption to Celia Clark

[6] National Trust [US] *Public Policy Weekly Bulletin* March 10, 2006, Volume 1, Issue No. 4

[7] Clark Celia 2000 *Vintage Ports or Deserted Dockyards; differing futures for naval heritage across Europe* The University of the West of England Working Paper No. 57 July 2000 Bristol

[8] Woodward Rachel 1998 Military Base Conversion and Rural Development ESRC Application, Centre for Rural Economy, University of Newcastle-upon-Tyne

[9] Albrecht, Kaldor M & Schméder G 1998 *Restructuring The Global Military Sector Volume II: The End of Military Fordism*

[10] Kaldor M and G. Schméder (1997) *The European Rupture: Defence Sector in Transition* Elgar, Cheltenham

[11] Broska 1998 'Conversion in a resource-reuse perspective' COST 10 Network workshop Bled Slovenia July

[12] Ministry of Defence 1998 *Strategic Defence Review* Cm 3999 TS0 London

[13] Chancellor of the Exchequer 1998 *Comprehensive Spending Review Modern Public Services for Britain* Csr Cm 4011

[14] Woodward Rachel 1998 *Rural Development and the Restructuring of the Defence Estate: A Preliminary Investigation* Research Report, Centre for Rural Economy, University of Newcastle upon Tyne

[15] Bonn International Centre for Conversion 2006

[16] Bonn International Centre for Conversion *Conversion Surveys of Global Disarmament, Demilitarization and Demobilization* of 1996, 1997, 2002 and 2003

[17] www.bicc.de/publications/brief/brief18/content.php

[18] Myrttinen Henry 2003 *Base Conversion in Central and Eastern Europe 1989-2003* Bonn International Center for Conversion Bonn

[19] Batchelor Peter, Cock Jacklyn & McKenzie Penny 2006 *Conversion in South Africa in 1990's: Defense Downsizing and Human Development Challenges BICC Brief 18* Bonn International Center for Conversion and Group for Environmental Monitoring, Johannesburg 61 pp

[20] DLO and Defence Estates n.d. *Project Prime Portsmouth Regeneration and Investment in the Maritime Estate* HM Naval Base Portsmouth

[21] Wildridge Graeme and Marshall Gavin Defence Estates 2005 Interview with Celia Clark Semaphore Tower 21 January 2005

[22] Hussain Amjad Cdre 2004 'Development Plan for Portsmouth Naval Base' 16 June *Navy News* Portsmouth http://www.navynews.co.uk /articles/2004.0406/000461601.asp

[23] Goodship Peter & Carpenter Lorraine 2005 Naval Base Property Trust Interviews with Celia Clark The Hard, Portsmouth 21 January 2005, 1 March 2005

[24] Haler Peter 2005 in Barry Jane 'A common touch. It is time for flat-owners to assert their newly won rights. Now Britain is only 50 years behind' Lease www.lease-advice.org *Evening Standard Homes & Property* 26 January 2005 p.10 London

[25] *The News* Portsmouth March 17 2005 p.8 Johnston Newspapers Plc. Portsmouth UK

[26] Maddox David 2005 "Tories will save ships from axe, says Howard" *The News* 30 March p.2 Johnston Newspapers Plc. Portsmouth UK

[27] Norwegian Defence Construction Service 1999 Letter to Celia Clark 9 December

[28] DLA Gordon 1997 Managing the changing political environment in urban waterfront redevelopment *Urban Studies* 1997 p 34; DLA Gordon 1999 'Implementing urban waterfront redevelopment in an historic context: a case study of the Boston Naval Shipyard p.925 *Journal of Ocean and Coastal Management 42 (1999)*

[29] Woodward Rachel 2004 *Military Geographies* Blackwell Publishing Oxford p.3

Remediation of former military sites

J. Grima & R. García-Delgado
Geological Survey of Spain, Spain

Abstract

The European Parliament's mandate to the Commission to develop a thematic strategy for soil protection highlights the need to adopt measures that prevent, limit or reduce the impact of human activities on soil. In Spain, the Royal Decree 9/2005 of 14th January fulfils the provisions of the Wastes Law 10/1998 of 21st April, and establishes a list of potentially soil contaminating activities and criteria and standards for declaring that sites are contaminated, subject to prior consultation with the Autonomous Communities. The Royal Decree shall not apply to publicly owned sites where military installations are located. Within a period of two years from the date on which it enters into effect, the Ministry of Defence shall approve, subject to previous acceptance by the Ministry of the Environment, a decontamination plan for such sites. As the Ministry of Defence owns more than 1% of the whole National territory, the selection of a case study is the first step in the definition of a practical methodology for demonstration of the economic viability of the environmental remediation. It involves estimating a range of values for the potential profit that could be obtained when the site is disaffected from use and introduced into the market.
Keywords: catalogue of contaminated properties, decontamination plan, defence facilities, polluting activities, soil pollution, study of economic viability.

1 Introduction

The number of contaminated sites in Western Europe (Prokop et al. [1]) varies between 300,000 and 1,500,000. These figures, in themselves indicative of the seriousness of the problem, also point to the serious ecological and legal consequences that result from an absence of standardized methodologies for identifying and characterizing contaminated soils. Indeed, the extent of the difference between the two figures is due precisely to a lack of uniformity in the

WIT Transactions on Ecology and the Environment, Vol 94, © 2006 WIT Press
www.witpress.com, ISSN 1743-3541 (on-line)
doi:10.2495/BF060201

criteria used in different countries for defining contaminated sites, quantifying acceptable risks and adopting instruments and characterization methodologies.

In Spain, the Wastes Law of 1998 identifies the polluter as liable (Ministry of Environment [2]). If the polluter is unknown, the landowner is responsible. The Ministry has the option to fund a remediation and later recover the costs. Costs can be recovered either directly or by the transfer of an appropriate portion of the remediated property (Lopez de Velasco [3]). Cost recovery is to take place over a 10-15 year period. The Wastes Law includes a mandate instructing the Spanish Government to approve and publish a list of potentially soil-contaminating activities, and establishes specific obligations that apply to the owners of activities and of properties where the specified activities are carried out or have been in the past

The Royal Decree 9/2005 of 14th January (Ministry of Presidency [4]) establishes a list of potentially soil contaminating activities and criteria and standards for declaring that sites are contaminated. It fulfils the provisions of the Wastes Law 10/1998 of 21st April, subject to prior consultation with the Autonomous Communities. The Royal Decree makes reference to the presence of dangerous chemical substances of human origin that are capable of altering the chemical, physical or biological characteristics of soil. They, therefore, constitute a risk that must be quantified in order to assess possible harm that may result to human health and the environment. Sites shall be declared contaminated by express decision if, in accordance with the standards set out in this Royal Decree, said risk is deemed unacceptable for human health and the environment.

The Contaminated Soils Decree shall not apply to publicly owned sites where military installations are located or where military activities are carried out. In the past, many military activities have been developed in industrial belts located in the surroundings of the cities. Such activities can be compared to industrial ones, and due to the growth of many cities, these sites have been absorbed and now are part of the reserves of land, required for the future development of the city.

Within a period of two years from the date on which this Royal Decree enters into effect, the Ministry of Defence shall approve, subject to previous acceptance by the Ministry of the Environment, a decontamination plan for such sites, which shall conform to the technical content of the Royal Decree. As a consequence, the Ministry of Defence is responsible for the design and preparation of a catalogue of contaminated properties to be remediated in the future. The Spanish Ministry of Defence is the first landowner in Spain, with an estimated 1% of the whole national territory. These figures are representative of the huge task to be accomplished.

2 Process for the remediation

The environmental remediation of contaminated properties, especially former military sites, requires a study of economic viability. Methodologies addressed in such a study have been tested and demonstrated in other countries, but application in Spain has to be done prior to starting any big scale remediation plan (Grima and García-Delgado [5]). The objective of the study is the definition of the steps for an effective environmental remediation work. The scope and performance of remediation actions shall be such as to guarantee that any remaining contamination translates into acceptable risk levels in relation to current and anticipated land uses. To illustrate the process, a property of the Defence Ministry has been chosen as a case study. The demonstration of the viability of the remediation from the economical and technical point of view is a prerequisite to initiate the process.

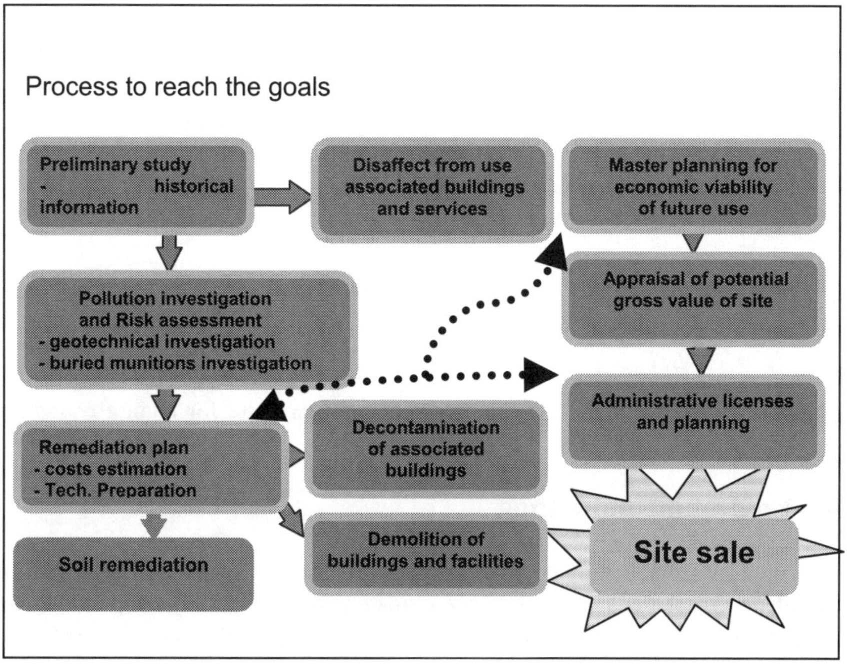

Figure 1: Main activities to be done for remediation of Defence Ministry properties.

3 Historical land use

At this stage, information about the historical land use of the site and its surroundings is to be collected. In addition, activities carried out, pollutants generated and accidents (e.g. spillages) that occurred during the useful life of the facilities must be investigated.

The former fire range studied is located in the municipality of Lorca, province of Murcia, approximately 45 km south of Murcia City. The property has an extension of 688,484 m^2. The estimated value of the property is 353,395.12 €. Little information exists about previous land use before it was selected as a firing range by the Spanish Army, but based on this information agricultural use seems to be the traditional use of the site from centuries past.

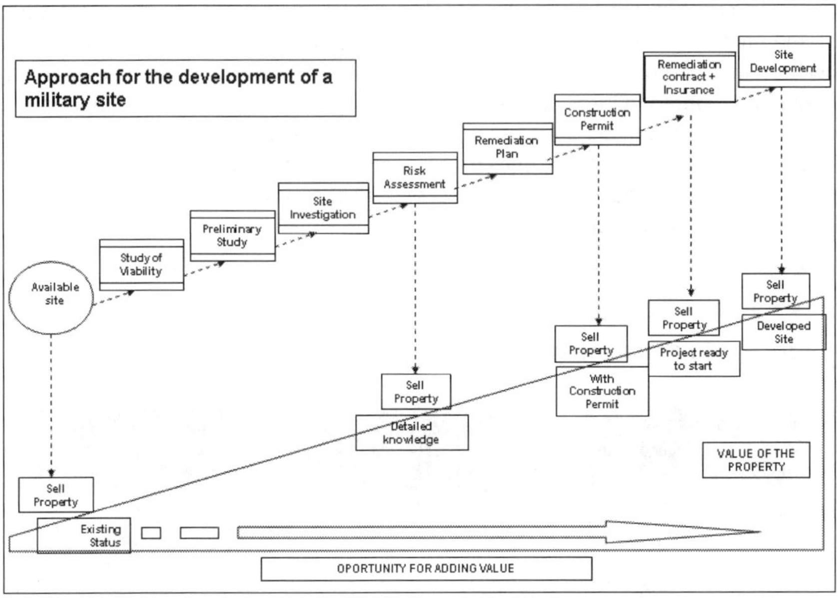

Figure 2: Steps in the development of a military site for adding economic value.

4 Environmental setting

The environmental setting comprises the factors that define the characteristics of the site, and include, at least, geology, climate (average monthly temperature and precipitation), risk of flooding (flood return periods, etc.), hydrology, hydrogeology and ecology (natural factors and human influences).

Generally speaking, groundwater is the most important pathway for migration of contaminants, so special attention must be paid to the definition of the site materials. The existence of aquifers must be determined as well as directions of flow, relationships between piezometric levels and recharge and discharge areas.

Moreover, information about springs, water wells, hydraulic parameters (transmisivity, storage coefficient, etc.), surface water and relationship between groundwater and surface water must be analyzed. The Commission proposal of Groundwater Directive COM(2003)550 specifies that the "good chemical status" of Groundwater shall be based on compliance to existing Community quality standards (nitrates, plant protection products and biocides) and on threshold values (i.e. environmental quality standards) to be established by Member States at the most appropriate level (local, regional or national). These threshold values will have to be set up for all pollutants which have characterised groundwater bodies as being at risk, pursuant to the analysis of pressures and impacts carried out under the article five of the Water Framework Directive. The agenda of the

Groundwater Daughter Directive proposal, as modified in the Council Common Position, implies that threshold values for relevant groundwater pollutants should be established on the basis of a common methodology before December 2008.

Figure 3: Aerial view of a former military site.

5 Potential contamination sources and risk assessment

The first objective is the review and evaluation of known and suspected contamination sources. Potentially polluting activities must be analyzed and evidence of buried munitions or contamination must be searched for.

Without prejudice to any subsequent development of requirements by the Autonomous Communities, the elements that shall be included in assessment of risks associated with contaminated sites are as follows:

- A detailed description of the focal points of contamination, identifying the contaminating substance(s) and determining significant concentration values (maximum measured, p95 or another duly justified statistic).
- A characterization of the textural properties and components of the soil.
- A description of the immediate physical environment for the purpose of identifying mechanisms that may transfer contaminants from focal points to potential receptors and relevant contamination exposure pathways for such receptors, including groundwater.
- An identification of potential receptors of contamination and an assessment of the characteristics or habits that determine their level of exposure to contamination. In the absence of other information concerning such characteristics or habits, this assessment may make use of the parameters utilized to define reference levels. The existence at the site in question or in adjacent areas of relevant ecological receptors shall also be taken into account.

- Identification of foreseeable exposure pathways and quantification of the exposure dose received. However, possible pathways may be added or eliminated based on the expert opinion of the technicians responsible for carrying out the assessment, subject to previous consultation with the Autonomous Community official in charge. For the quantification of exposure dose, the expressions used to develop reference levels or similar methods deemed appropriate by the Autonomous Community officials in charge may be employed.
- A toxicity value and justification of its choice for each of the relevant contaminants identified.
- Quantification of risk. In cases where contaminants with the same action mechanism are found together at a single site, the combined risk that they pose shall be considered.
- An analysis of the uncertainties associated with the risk assessment carried out, including appropriate conclusions regarding the validity and reliability of the results of the assessment.

The degree of detail with which these points are to be covered shall be established by the competent body of the Autonomous Community in question, based on reasonable criteria and taking into account the specific circumstances that apply in each case.

6 Study of economic viability

The first phase to consider in a study of economic viability is the appraisal of the potential gross value of the site. Then, the different potential costs associated with the development of the site must be calculated and the ecological protection requirements must be included.
Among the potential costs, the following ones must be considered when making calculations: General communications (railway and access), protection of surface waters (and channels), general infrastructure (electricity, sewerage, water supply, etc.) and, finally, remediation costs.

Sampling will determine the degree of remediation needed but, based on similar experiences, a conservative estimation of remediation costs at Carraclaca pilot site provides a figure between one hundred and twenty thousand euros (120,000 €) and one million, three hundred and fifty thousand euros (1,350,000 €). Therefore, the potential net benefit for the owner of the site varies between some five million, seven hundred thousand euros (5,777,576 €) and seven million euros (7,007,576 €), compared to the potential gross value of the site at the beginning of the process, of about three hundred and fifty thousand euros (353,395 €).

On the other hand, the Ministry of Defence has established in the last years a very efficient model of compensations (or spin-offs). The return of a certain established percentage of the military investment has proved to be a very important source of funding. The qualitative impact of spin-offs has had impact on sectors and activities of major interest from the economic, technological and defence point of view. Priority is given to sensible technologies in acquired equipment and formation of personnel and establishment of centres of excellence for research and development from the knowledge transferred.

Figures in euros	CASE A	CASE B
REVENUE		
Potential Gross Value	**7,692,826**	**7,692,826**
COSTS		
INVESTIGATION		
Geophysics	38,000	38,000
Sampling	75,000	75,000
REMEDIATION		
Project Management	90,000	90,000
Explosives Cleaning	362,250	362,250
Structures Demolition	90,000	90,000
Contaminated Soil Removing	120,000	1,350,000
SUBTOTAL	**775,250**	**2,005,250**
NET BENEFIT (DEFENCE)	**6,917,576**	**5,687,576**
Actual value	**353,395**	

Figure 4: Summary of investigation and remediation costs at Carraclaca pilot site.

7 Conclusions and recommendations

The Wastes Law of April (1998) determines the existence of Contaminated Soil and the different responsibilities of landowners. It has implications from the legal and economic point of view.

The provisions of the law have been developed by means of the Royal Decree 9/2005 of 14th January. The manner in which a record is to be created in the Land Register and the declarations made by property owners carrying out potentially contaminating activities are defined in the Decree, giving effect to the legal provision by virtue of which a marginal note is to be recorded, its content, effects, duration and corresponding requirements for cancellation. Within a period of two years from the date on which this Royal Decree enters into effect, the Ministry of Defence shall approve, subject to previous acceptance by the Ministry of the Environment, a decontamination plan for such sites, which shall conform to the technical content of the Royal Decree.

Due to the different land uses and remediation technologies, calculation for different land use scenarios should be consider in order for better return on capital investment.

The rehabilitation of a contaminated site must be undertaken in phases, according to a sequential program. It is not feasible to make precise estimations of remediation costs before finishing site investigation, so there is a real risk of underestimating the total remediation cost in early stages of the process.

According to the study of economic viability conducted on the Carraclaca pilot site, the application of the proposed approach can optimize the benefits related to the disaffection of a military site and introduction into the market. National databases can provide very useful information on natural characteristics of investigated sites.

References

[1] Prokop, G., Schamann, M. & Edelgaard, I. Management of Contaminated Sites in Europe. European Environment Agency. Topic Report Nº 13, pp 8-16, 1999.

[2] Ministry of Environment. Wastes Law. Official Gazette of the Spanish State, Nº 96, pp 13,372 – 13,384, 1998

[3] Lopez de Velasco, J. Ministry of Environment. Personal Communication, pp 12-14, 1999.

[4] Ministry of Presidency. Royal Decree of Contaminated Sites and potentially contaminating activities. Official Gazette of the Spanish State, Nº 15, pp 1,833 – 1,843, 2005

[5] Grima, J. & García-Delgado, R. Study of economic viability for remediation of former military ranges. Internal communication, pp 1-51, 2005

Section 8
Modelling and assessment

Green economics as a tool for prioritising brownfield rehabilitation in Greece

D. Kaliampakos, D. Damigos & T. Karachaliou
Laboratory of Mining and Environmental Technology,
School of Mining Engineering and Metallurgy,
National Technical University of Athens, Greece

Abstract

Given the limited resources, an effective way of prioritizing brownfield rehabilitation could be according to the total costs and benefits produced by each site. Aside from the potential return from a profitable new use, brownfield rehabilitation benefits also come from contamination and stigma elimination. The latter, however, are not directly incorporated in the market, and therefore a comparable value is missing. Environmental valuation methods could monetize environmental enhancement in order to attain the comparison of benefits with remediation costs, on the same basis. Towards this direction, an interactive tool is currently being developed, namely the Greek Environmental Valuation Database (GEVAD), on the grounds of the benefit transfer method. This tool can facilitate the valuation of brownfield external costs and benefits, improving the justification of rehabilitation funds.
Keywords: brownfield, Greece, case studies, environmental economics, cost benefit analysis, benefit transfer.

1 Introduction

Over the past decades, almost all of the developed countries have restructured their traditional heavy manufacturing industries, causing older sites to be abandoned. This is the result of stricter environmental regulations, changing markets and innovative production technology. In turn, communities have experienced downsizings, shutdowns and relocations, leaving underused or vacant industrial properties behind. Considerable effort has been made, in the USA and in many European Countries, to understand the nature of the problem

WIT Transactions on Ecology and the Environment, Vol 94, © 2006 WIT Press
www.witpress.com, ISSN 1743-3541 (on-line)
doi:10.2495/BF060211

and provide tools to resolve it. Greece has to face, too, the environmental and socioeconomic implications caused by brownfield land. Until now, though, there is no evidence of systematic, concerted actions towards brownfield redevelopment. This may be due to the fact that there is no strong public awareness of the situation, while the Greek government does not support adequately the voluntary initiatives with financial incentives.

According to a recent study of the Laboratory of Mining and Environmental Technology of the National Technical University of Athens [1], the list of some notable brownfield sites that have been recorded in Athens includes Factory "Fix", Gasworks "Gazi", French Lavrio Mines Company, Silk-mill "Athinais", Factory "Thermis", Factory "Phoivos", Former National Printing Office, Public Tobacco Factory, Steam Power Station of Neo Faliro, Factory-Engine Room of OSE, Printing Office "Empros", Greek Company of Chemical Products and Fertilizers, etc.

The number of abandoned or underused industrial sites, posing threats to human health and the environment, is quite big and is still increasing, yet many of them offer the grounds for affordable remediation, if needed, and further redevelopment. The cases described below, are indicative of their significant potential, even when the conditions seem initially unfavorable.

2 Brownfields in Athens - success stories

2.1 Lavrio Technological and Cultural Park

Lavrio Technological and Cultural Park is one of the most important projects related to brownfield regeneration in Greece. It is located within the metropolitan area of Athens, on the north side of the city of Lavrio. The site holds a significant mining past dating back to ancient times, while in the recent history accommodated the French Lavrio Mines Company, the most significant industry at the time. The mining company was established by J. B. Serpieri, in 1875, in order to exploit the ancient slags and extract the silver-containing lead ores and was responsible for the foundation of a 10,000 residents' company-town and multiple technological innovations.

In 1990 deindustrialisation trends in specific areas in Greece forced the industrial facility to shut down its units. The remains were buildings of great architectural and aesthetical value. The environmental site assessment, however, showed the presence of many heavy metals in the soil, which was characterized as unusually heavy contaminated. In order to deal with the contamination and, at the same time, provide for a feasible activity, the National Technical University of Athens undertook the redevelopment of the old French Mine Company complex as a Cultural and Technological Park. The remediation plan for the site includes excavation, transfer and disposal of the contaminated land and construction of an underground space used as a hazardous waste storage facility. After having partially restored the site and ensuring that the risk to human health has been eliminated, the Park started to operate [2]. To date, fifteen buildings have been restored, some of which are already leased, whereas the rest of them are intended to accommodate administrative and cultural services. The restoration of the buildings was guided by an attempt to turn into account their original functionality, maintaining their industrial character.

2.2 Former factory "Gazi" - "Technopolis"

The former gasworks factory, known as "Gazi", was established in 1857, under the directorship of a French company in order to illuminate the city of Athens. It was when the industrial revolution was just starting to make its appearance within the newly founded Greek State. The siting of the industrial complex was based upon its central position. The development of the surroundings, later on, conduced to the recognition of the area as one of the most significant industrial places until nowadays. In 1887, the need for town gas expanded. In addition to lighting the streets, homes and industries were added to the list of gas consumers. During the years 1920-1952 German technology was introduced to improve the quality of gas. The last phase of operation (1952-1984) marked the gasworks decay. In the 1960s, the gasworks did not function at full capacity, as the specific form of energy was considered 'outdated'. In 1983 the gas production process from coal was suspended. Gasworks operation was considered anachronistic, raising intense reactions from the public, not only due to its environmental burden, but also due to its vital location in the city core, near Acropolis. Thus, in 1984, the factory's function was ceased.

Today, the site has been redeveloped as a cultural, modern multipurpose center, known as "Technopolis". The rehabilitation of the buildings was attained, abiding by two basic principles: conservation of the industrial past authenticity along with functionality. The result consists of a living Industrial Archeology and Architecture Museum, with the ability to host important events of science and the arts.

2.3 Former silk-mill "Athinais"

Silk-mill "Athinais" was established in 1920, in a location facing the Acropolis, near the Ancient Agora of Athens and operated for two decades, producing natural silk of the finest quality. It was recognised as one of the city's modern-age landmarks. During World War II, the factory suspended operation to become an air-raid shelter. Following the city's liberation, "Athinais" started running again, but had to face post-war technological advancements, which had begun to revolutionize the industry. Finally, in the mid 50's, unable to march with the times and compete with rayon, which gradually replaced natural silk, the old factory ceased operation and was merely used as a storage facility. Today, the 6,500 m^2 old silk factory has been converted into a modern multi-purpose centre that hosts a wide range of business initiatives and cultural activities, such as conferences, business meetings and exhibitions. The renovation project was designed in a way to preserve the building's industrial character and accentuate its distinctive architectural features.

3 Reconsidering brownfield prioritization

Despite the existing potential, there are many obstacles opposing to brownfield redevelopment. The most important one is the lack of the necessary resources to restore all the sites, meaning that, inevitably, a choice is required. In USA, each recorded brownfield is classified in the National Priorities List, which is intended to determine which sites warrant immediate investigation and possibly remediation. The ranking of the hazardous sites is founded on a screening tool, called the Hazard Ranking System (HRS), which assigns numerical values to factors related to human health and ecological risk, depending on the site's conditions. As far as Greece is concerned, the decision-making seems to be mostly guided by the site's history and location, its special industrial architecture and some personal memories and feelings rather than a systematic method focused on the optimization of the restoration funds or the elimination of the environmental burdens.

Brownfield evaluation, based on risk assessment methodology is justified by the fact that these sites may be related to adverse environmental impacts. And for that it is a sound first step. However, this method is inadequate, in cases where the environmental problems are alike but the consequences differ, depending on the number of people or species that are affected. In that case, while the cost remains the same, there is a difference in the social benefits deriving from remediation, which cannot be estimated through the risk assessment approach. Thus, in order to optimize the allocation of the available funds, the remediation and redevelopment benefits gained should be quantified and compared to the costs required.

Nowadays, there is an extensive literature indicating that contaminated sites are affected from the so-called stigma-related damages. Stigma generally refers to the loss in value of a property that has suffered some type of contamination. This loss is not attributed to remediation costs but on perceived risks associated with the contamination. In many situations, it is documented that stigma extends beyond the boundaries of the site, reducing the value of nearby residential, as well as commercial and industrial properties, at least temporarily. In general, it is well established that value reduction is associated with the distance from the contaminated site. For example, Reichert [3] found that properties within a distance of 700 m from an Ohio Superfund toxic waste site lost approximately 15% of their value, while losses were negligible 2 km away from the site. Cumulative losses attributed to this hazardous site were nearly 11 million USD (1994). The Office of Emergency and Remedial Response [4] found that the cumulative property value loss from 1,213 hazardous sites amounted to approximately 750 million USD (1996) at the one-mile threshold, and 2.1 billion USD (1996) at the two-mile threshold. Although in 1997, the OERR dismissed the results of the previous study [5], it was suggested that price effects range from 2 to 8% of the original property value within a range up to 3 to 7 miles. Similar results are reported from several studies [e.g. 6–9].

The following section outlines some fundamental definitions and principles of environmental economics, along with the basic techniques, in order to provide the theoretical and practical background of environmental valuation.

4 Environmental valuation

Although environmental assets are valuable to the society, there is no market mechanism assigning prices to environmental goods and services. In order to incorporate cost - benefit analysis in the decision making process, as in the case of brownfields, it is necessary to provide a common measure of comparison between benefits and costs, using monetary values. Environmental valuation is attained by estimating the effect of a change in environmental conditions on social welfare, which is called the total economic value (TEV) of the change. Practically, this is based on people's willingness to pay (WTP) an amount of money in order to avoid an environmental degradation or their willingness to accept (WTA) a compensation in order to suffer the environmental impacts incurred [10–13]. TEV can be divided into *use values* and *non-use values* reflecting the fact that people also value resources for moral reasons and bequest reasons, unrelated to current or future personal use of the good.

There are two kinds of original environmental valuation techniques: the revealed preference methods that estimate the people's actual behavior and the stated preference techniques that estimate the intended behavior. Revealed preference methods include Travel Cost Method (TCM), which is mainly used to measure the demand for recreational activities through time and travel cost expenses to access a site and Hedonic Pricing Method (HPM), which is used to estimate the economic value of environmental assets that directly affect market prices, such as air and water pollution, noise, proximity to recreational sites etc. HPM is usually used to measure variations in the house market, reflecting the value of the local environmental quality. Finally, Contingent Valuation (CVM) is referred as a stated preference method, because it is an experimental technique that directly asks people what they would be willing to pay or to accept in a hypothetical situation, instead of inferring values of actual choices. However CVM is the most popular method because it can estimate the total economic value, both use and non-use. CVM has been in use for over 35 years and there are more than 2,000 studies dealing with the topic [14].

Despite their recommended use, original environmental valuation methods are a long lasting and costly procedure, while there is usually limited time and resources. As a result, a method based on secondary data, known as Benefit Transfer Method (BTM), is often applied. BTM makes use of the results of one or more primary studies to infer the values for the environmental impacts or benefits for the site of concern. The site of the previous research is usually called the "study site" while the site to which the benefit estimate is transferred is called the "policy site". BTM is a useful alternative when original research is not possible or there is no need for high accuracy. The method has been criticized for not producing reliable values, thus, in most of the cases the inaccurateness is due to the imprecision and the deficiency of the original study, the difference in the research and statistical methods applied or the instability of data over time. In order to reduce the margin of error of a benefit transfer application, several conditions should be met [15, 16]. Studies transferred must be based on adequate data, sound economic method, and correct empirical technique [17]. The study site and the primary policy site should exhibit similar environmental resource and change in the quality (or quantity), similar markets for the sites, unless there is enough usable information provided by the study on own and substitute prices, similarity of demographic characteristics between the two populations and their cultural aspects and similar extent and magnitude of the population that will be affected by the expected site or resource impacts.

There are two benefit transfer techniques: (a) value transfer and (b) function transfer. Value transfers consist in the transfer of a single estimate from a study site, or a measure of central tendency of the estimates from several study sites. Function transfers consist in the transfer of a benefit or demand function from a study site that is then adopted to fit the special elements of the policy site [e.g. 16]. When transferring values, spatial and temporal adjustments should be made, regarding differences of income, price level and time.

5 Greek Environmental Valuation Database (GEVAD)

Reaching a satisfactory degree of reliability in benefit transfer application requires having access to a large range of original studies in order to find the most appropriate ones, which have similar characteristics with the study site. So far, two tools are well known and widely used: the EVRI (Environmental Valuation Reference Inventory) and the ENVALUE databases. EVRI is a web-based database that was developed by Environment Canada in collaboration with a number of experts and organizations. ENVALUE was constructed by New South Wales Environment Protection Authority (NSW EPA) and was last updated in October 1998. These two tools can be searched by environmental attribute valued, valuation technique used, location etc, so as to identify the most relevant primary studies. Despite the fact that EVRI and ENVALUE are globally accessed for retrieving field valuation studies, they both have weaknesses as far as applications in Europe are concerned. According to Navrud and Vågnes [18], only 56 out of the more than 650 European Valuation studies were recorded in EVRI, which corresponds to less than 10%.

The European Union's motivation towards sustainable development and the increasing need to include environmental valuation in the decision-making processes emerged the necessity to develop a similar tool focused mainly on recent European studies. A relative project, which is co-funded by the European Social Fund and Greek National Resources, is being carried out by the Laboratory of Mining and Environmental Technology of the National Technical University of Athens.

More specific, the scope of the appointed project is the development of a user-friendly, interactive "environmental valuation" tool, in compliance with the institutional and research context of the international scientific community. The tool will provide the necessary data to value environmental impacts of industrial activities in Greece, as well as in other European countries, based on the BTM. The prescribed actions of this project include literature review of a large number of studies, codification and adjustment to the particular characteristics of European space using suitable techniques and development of a web-based software to guide the users. The GEVAD tool (Greek Environmental Valuation Database) is intended to be completed till the end of 2006. By that time, approximately 400 studies will have been registered. To date, 1200 studies have been tracked and 1017 from them are selected to be reviewed in the next phase, focusing on the ones that are spatially and temporally more relevant to Greece and Europe, in general.

Indicatory of the emphasis given on the most recent research is the fact that 42.9% of the studies selected were conducted in the last five years and almost 75% in the last ten years. Table 1 lists the classification of the studies according to their date.

Table 1: Studies classification according to their date.

Year	Portion %
1970-1974	0.4%
1975-1979	2.0%
1980-1984	4.5%
1985-1989	5.0%
1990-1994	14.2%
1995-1999	31.0%
2000-2005	42.9%

Additionally, 57.0% of the recorded studies concern Europe, filling the gaps of the databases mentioned (Figure 1).

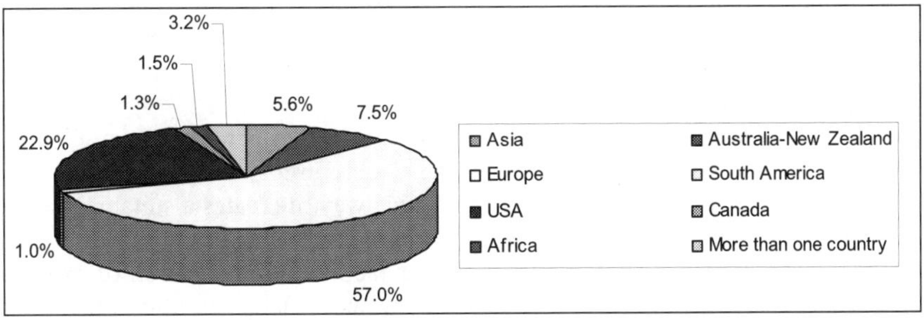

Figure 1: Studies classification according to geographical area.

Studies are classified in GEVAD according to the environmental asset, good or service, which is valued. Amenity, water and air contamination along with landscape degradation seems to be the most interesting attributes to be estimated in monetary terms (Figure 2).

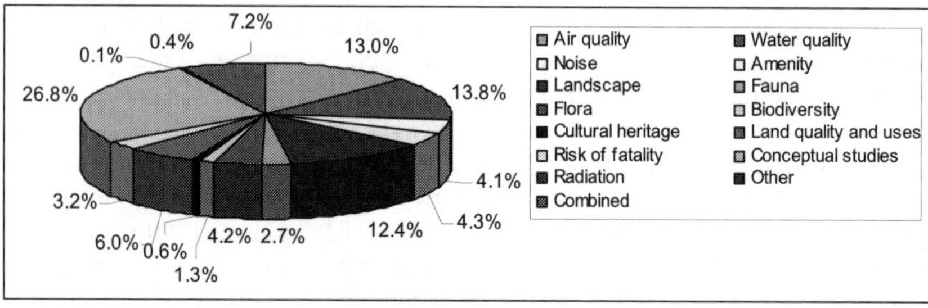

Figure 2: Studies classification according to the environmental asset that is valued.

6 Conclusions

Brownfield regeneration is an established trend in the modern world. However, the lack of the necessary funds for dealing totally with the problem, calls for a clear and sound way for prioritizing the various cases, in terms of emergency and net social benefits gained.

So far, prioritization of these sites is accomplished by means of risk assessment techniques. Nevertheless, brownfield redevelopment is, first of all, an economic puzzle. Towards this direction, the application of environmental valuation methods could be beneficial, in order to come up with more fair choices. The monetization of environmental costs and benefits reveals hidden economic aspects and provides the means to evaluate a redevelopment project not only according to its financial profitability but also on its performance towards sustainable development.

Although primary studies are always the "first-best" option, Benefit Transfer Method is well recognized as a cost-effective alternative, when original study of the problem is not possible due to budget constrains or time limitations. In any case, values should be transferred cautiously, taking into account that the results are affected by several parameters (e.g. socioeconomic characteristics of the "policy" and "study" sites, stability of data over time, since the primary studies occurred at different points in time, etc.).

GEVAD, which stands for "Greek Environmental Valuation Database", aims at expanding the coverage provided by the widely used EVRI and ENVALUE databases to recent EU valuation studies. In this way, it creates a suitable pool of data that will encourage the application of benefit transfer in Greece and other European countries. This attempt will further facilitate the valuation of external costs and benefits of several types of projects and activities, including brownfield. Hence, it can be used as a tool for the prioritization of the hazardous sites, on an economic basis, in order to appease the burdens involved in decision-making processes.

Acknowledgements

The GEVAD Project is co-funded by the European Social Fund (75%) and National Resources (25%) - Operational Programme for Educational and Vocational Training (EPEAEK II)- PYTHAGORAS.

References

[1] Kavallari E. *Old industrial areas (brownfields) redevelopment- the case of Attica*, dissertation, NTUA, 2005.

[2] Karachaliou T., Kaliampakos D., *Redeveloping derelict urban space, the case study of Lavrio, Greece*, proceedings of the International Conference on Managing Urban Land, CABERNET 2005.

[3] Reichert, Alan K. *Impact of a Toxic Waste Superfund Site on Property Values*. Appraisal Journal, 65, 4, pp. 381-92, 1997.

[4] Office of Emergency and Remedial Response. *Estimating Property Value Impacts From Superfund Site Cleanup*, 1996.

[5] Office of Emergency and Remedial Response. *Plan for Socioeconomic Analysis*, Working Draft, 1997.

[6] Thayer, M., Albers, H., Rahmatian, M. *The benefits of reducing exposure to waste disposal sites: a hedonic housing value approach*, J. Real Estate Res., 7, 265–282, 1992.

[7] Elliot-Jones, M. *Stigma in light of recent cases*, Nat. Res. Enviro. 56–59, 1996.

[8] Farber, S. *Undesirable facilities and property values: a summary of empirical studies*, Ecol. Econom. 24, 1–14, 1998.

[9] McCluskey, J.J. and Rausser, G.C. *Estimation of perceived risk and its effect on property values*, Land Econom. 77, 42–55, 2001.

[10] Pearce, D. and Turner, R.K. *Economics of natural resources and the environment*. Harvester Wheatsheaf, Hertfordshire, U.K., 1990.

[11] Johansson, P-V. *Cost-Benefit Analysis of Environmental Change*. Cambridge University Press, Cambridge, 1993.

[12] Turner, R.K., Pearce, D. and Bateman I. *Environmental economics: An elementary introduction*. Harvester Wheatsheaf, Hertfordshire, U.K., 1994.

[13] Kula, E. *Economics of Natural Resources, the Environment and Policies*. Chapman and Hall, London, U.K., 1994.

[14] Carson, R.T., Wright, J., Alberini, A., Carson N., Flores, N. A Bibliography of Contingent Valuation Studies and Papers. *Natural Resource Damage Assessment*, Inc., La Jolla, January, 1995.

[15] Desvousges, W.H., Naughton, M.C., Parsons, G.R. *Benefit transfer: Conceptual problems in estimating water quality benefits using existing studies*. Water Resources Research. 28(3):675-683, 1992.

[16] Rosenberger, R.S. Loomis, J.B. *Benefit transfer of outdoor recreation use values: A technical document supporting the Forest Service Strategic Plan*, (2000 revision). Gen. Tech. Rep. RMRS-GTR-72. Fort Collins, CO:

U.S. Department of Agriculture, Forest Service, Rocky Mountain Research Station, 2001.

[17] Freeman, M., III. *On the tactics of benefit estimation under Executive Order 12291*. In: Smith, V. Kerry, ed. Environmental policy under Reagan's Executive Order: The role of benefit-cost analysis. Chapel Hill, NC: The University of North Carolina Press, chapter 6, 1984.

[18] Navrud, S. and Vågnes, M. *Assessment of Environmental Valuation Reference Inventory (EVRI) and the Expansion of Its Coverage to the EU*, 2000, http://europa.eu.int/comm/environment/enveco/studies2.htm#24

A model for redeveloping complex, highly contaminated sites — the Industri-plex Site in Woburn, Massachusetts

C. N. Brooks
Greenfield Environmental Trust Group, Inc. (GETG),
Resources for Responsible Site Management, Inc. (RRSM),
The Industri-plex Superfund Site Custodial Trust (U.S. Federal District Court, District of Massachusetts), USA

Abstract

This paper will describe how one of the most contaminated waste sites in the United States was transformed into a productive commercial center to become an international model for brownfields reclamation.

The Industri-plex Federal Superfund Site in Woburn, Massachusetts, is the fifth most contaminated waste site in the country, according to the US Environmental Protection Agency (USEPA). More than a century of heavy industry left a legacy of contamination, including metals in the soil, sediment and groundwater, and piles of odor-producing animal hides. Woburn is the home of two federal superfund sites—Industri-plex and the notorious Wells G&H Site, where contamination in two drinking water wells was blamed for a cancer cluster. After a highly publicized lawsuit and release of Jonathan Harr's best selling book, *A Civil Action*, local property values plummeted, community trust was shattered and Woburn became nationally stigmatized.

Today, however, this former industrial wasteland has been reborn and is now home to US Fortune 100 giants Raytheon Corporation, Marriott Hotels and Target, which occupy a 900,000 square foot high-end office park and hotel campus and a 200,000 square foot retail center. A state-of-the-art regional transportation center provides train, bus and carpool services to thousands of regional commuters. More than the 4,000 new jobs and millions in new taxes that have been created, Industri-plex is a source of new-found hope and pride.

This paper details the strategies that led to redevelopment success, including the unprecedented use of an independent, third-party redevelopment advocate and trustee, the "Industri-plex Custodial Trust," to catalyze and implement reuse.

Winner of the prestigious National Phoenix Award for excellence in brownfield redevelopment, Industri-plex offers an innovative model with broad application to many sites, especially those gridlocked by complex contamination, property issues and divergent stakeholder interests.
Keywords: superfund, stigma, risk, liability, community, USEPA, PRPs, remedy, institutional controls, Custodial Trust, trust, redevelopment.

1 Introduction

1.1 Site history

The Industri-plex Site ("Industri-plex") in Woburn, Massachusetts was one of the first properties to be placed on the USEPA Superfund National Priorities List ("NPL") of the ±1,200 most hazardous sites. Originally ranked number 3, today the 245-acre Industri-plex is the fifth most contaminated waste site in the US.

The City of Woburn, Massachusetts is situated approximately twelve miles north of Boston along the busiest interstate highway interchange in the state. Its strategic location along the Middlesex Canal, the Boston and Maine Railroad and the Aberjona River helped transform Woburn during the 1800s into the leather producing capital of the US and cradle of the American industrial revolution.

From 1853 to 1969 Industri-plex hosted pesticide, munitions and leather-related manufacturing, culminating in the production of glue from animal carcasses that were a by-product of the tanning industry. Tri-sected by the old Boston and Maine Railroad that once serviced industrial operations, today the Site abuts a major north-south interstate highway that was built just a decade before manufacturing ceased in the 1960s. The property was sold in 1969 to a real estate developer who began constructing roads and subdividing parcels for an industrial park. Then, in 1979, after the developer was found excavating and stockpiling buried wastes, significantly aggravating the existing contamination, government regulators brought site development to a halt.

1.2 Site contamination and the USEPA-selected remedy

A century of intensive industrial use left this 245-acre tract with extensive contamination. The major contaminants included: arsenic, chromium and lead in the soils and sediments; benzene and toluene in the groundwater; and four large piles of odor-producing animal hides.

To prevent exposure to contaminants by dermal contact, ingestion or inhalation, USEPA selected a remedy that required construction of permeable caps and covers over a 110-acre area of the Site. The minimum USEPA-approved "Basic Soil Cover" is comprised of a geotextile membrane covered by a layer of clean, vegetated fill. While protective of public health, the Basic Soil Cover precludes use of the property for anything but limited pedestrian access. Therefore, USEPA also allowed "Alternate Covers," such as asphalt or crushed gravel, to permit industrial or commercial use of the covered areas. Also, portions of the Site were already being used for light industry because the bankrupt, recalcitrant developer was able to subdivide and sell a number of parcels before his development was stopped. Consequently, USEPA chose to treat all existing buildings, parking lots and paved areas as "Equivalent Covers or Caps," thus allowing them to remain in place.

USEPA required use of a gas collection and treatment system to capture the hydrogen sulfide gas from one of the Hide Piles that was still decomposing and an interim groundwater remedy to treat the benzene and toluene.

The sheer volume of contaminated soils at Industri-plex (in excess of 100 metric tons for arsenic alone [1]) practically and economically precluded their removal and disposal off-site. Because the remedy entailed leaving the contamination on-Site, USEPA required a mechanism to ensure that the caps and covers would be maintained in perpetuity. Hence the remedy included "Institutional Controls" —enforceable land use restrictions that remain attached to the land in perpetuity. The final draft Institutional Controls take the form of a grant of easement and restriction to the USEPA to be recorded on title to all Site properties. They rely on deed restrictions, restrictive covenants and easements to restrict land use, and they are enforced by the USEPA to ensure long-term compliance. The risk-bask corrective action (RBCA) selected for the Site inherently restricted future use to commercial or industrial activities.

The agency recognized that metals from Industri-plex, particularly arsenic, were likely entering the groundwater and Aberjona River, although the final clean-up plan for Aberjona River was not issued until 2005.

1.3 Industri-plex Clean-up Settlement (Consent Decree) and creation of the Custodial Trust

In 1989 the USEPA reached agreement on the Site clean-up plan with twenty-two "potentially responsible parties" (or PRPs). The PRPs included current and former property owners, two large chemical companies that inherited the majority of liability for the cleanup, and the then-bankrupt real estate developer who had substantially aggravated the contamination. The clean-up agreement between USEPA and the PRPs was set forth in a 1989 federal district court decree. The so-called Industri-plex Consent Decree established two separate trusts—a Remedial Trust and a Custodial Trust. Representing the PRPs, the Remedial Trust was responsible for funding, designing, constructing and maintaining the Site remedy under USEPA oversight. Considering the PRP-developer's legacy of recalcitrance and insolvency, USEPA allowed the developer to contribute his property in exchange for relief from liability for the clean-up. Thus a Custodial Trust was created to take title to the bankrupt developer's property.

The Industri-plex Consent Decree assigned a pivotal role to the Custodial Trust in the redevelopment process, including responsibility for: (i) receiving, holding, managing and securing the insolvent developer's property; (ii) cooperating with the Remedial Trust and USEPA on Site remediation; and (iii) with the prior approval of USEPA and the two major PRPs, selling all salable land and distributing the sale proceeds to the three designated beneficiaries of the Custodial Trust—the City of Woburn, the Remedial Trust (PRPs) and the USEPA. The Trustee was able to take title to the Industri-plex property because USEPA agreed to protect the Custodial Trust from liability under superfund.

1.4 A case of civil action

The discovery of contamination at Industri-plex coincided with the detection of contaminants in two Woburn drinking water wells—Well G and Well H—

located on the Aberjona River south of the Site. The contaminated water was blamed for a childhood leukemia cluster that struck 28 children. Some of the Woburn families whose children died of leukemia filed a $400 million class action law suit against three companies believed responsible for the contamination. The widely publicized toxic tort trials stigmatized Woburn, depressed local property values, polarized the community and shattered Woburn's century-long relationship with industry. After the 1995 publication of Jonathan Harr's best seller, *A Civil Action* [1], which documents the litigation, and Hollywood's subsequent release of a major motion picture starring John Travolta, Woburn's stigma became a national phenomenon. Beyond the human tragedy, Woburn's trust in government and industry was destroyed.

2 The redevelopment process and results

2.1 Community partnering on redevelopment and infrastructure

2.1.1 Building an alliance with the community

In a climate fraught by mistrust and betrayal, the Custodial Trust pursued an alliance with the community. Clearly any reuse of the Site required community consent because land-use in the US is always governed at the local level. At Industri-plex the community's active participation was critical to overcoming the challenges posed by extensive contamination and a legacy of mistrust and market stigma. Working with local officials, the Custodial Trust undertook a series of environmental, land use and market studies to evaluate reuse options in light of the opportunities and constraints posed by the contamination and the remedy, existing and potential infrastructure, and other site characteristics. Soon local officials began to view the Site as an asset, rather than the liability it had come to represent.

2.1.2 Infrastructure: where redevelopment begins

Initial efforts focused on infrastructure, which is the foundation for any well-planned development. At Industri-plex, the Custodial Trust sought to improve and leverage existing infrastructure, such as the abutting highway, existing rail line, and other attributes that made the Site attractive for use in the first place. In partnership with the City of Woburn, the Custodial Trust led complex negotiations with three state transportation agencies—the Massachusetts Port Authority, the Massachusetts Bay Transportation Authority and the Massachusetts Highway Department (collectively, the "Transportation Agencies")—which resulted in a series of public-private agreements, and ultimately, construction of three major transportation projects at the Site: (i) a new interstate highway interchange providing direct highway access to the Site; (ii) a state-of-the-art intermodal, regional transportation center (RTC) that offers commuter rail, airport shuttlebus, van pool and park-and-ride services for 2,500 cars; and (iii) new public roads to reach landlocked areas. Under these public-private agreements the Custodial Trust donated the land, financed the acquisition of non-Custodial Trust property, and funded final design of the interchange; in exchange, the Transportation Agencies funded and constructed these projects.

2.1.3 Integrating re-use into the remedy and public-private agreements

To take advantage of the railroad that trisected Industri-plex, the 30-acre RTC was sited atop the highly contaminated former manufacturing plant parcel that abutted the railroad. The Transportation Agencies funded a more intensive "Alternate Cover" (up to 4-1/2 times thicker than the Basic Soil Cover) to enable final construction of the Central Passenger Station, underground utility installation, and potential expansion to a 5,000-car facility without coming into contact with contaminants. The RTC "Alternate Cover" operates as the remedy.

Further integrating future use into the remedy, new public roads were also constructed as the Alternate Cover for contaminated soils. The Custodial Trust built "clean fill" corridors so that underground utilities could be installed and maintained in the future without coming into contact with contaminated soils.

All told, the Custodial Trust helped secure almost $50 million in public capital for these infrastructure projects. That level of public investment not only created significant property value, it also sent a badly needed message to a risk-adverse private sector still struggling with concerns about environmental liability and market stigma. The state Transportation Agencies' decision to purchase, develop and occupy some of the most contaminated areas of the Site made a compelling case for the efficacy and safety of redeveloping Industri-plex.

2.1.4 Honoring the past and looking to the future

Determined to close a painful chapter in Woburn's history, thc Custodial Trust worked with the community, the state legislature and the Transportation Agencies to dedicate the RTC to Jimmy Anderson, one of the first Woburn children to die of leukemia. Jimmy was the son of Anne Anderson, whose personal testimony before a grid-locked US Congress was instrumental in securing passage of the superfund laws in the US. Today the Anderson RTC symbolizes Woburn's re-birth and a powerful reminder of its past.

2.2 Marketing and selling contaminated property

By the mid-1990s the remediation and infrastructure improvements were underway, freeing the Custodial Trust to begin the marketing and sale of its property. In collaboration with the City of Woburn, the Custodial Trust pursued a host of local zoning and permitting initiatives to facilitate the City's desired reuse and realize the highest and best use of the property for the benefit of its three beneficiaries. Then, capitalizing on the Site's emerging infrastructure and a regional market shortage for well-situated "box retail" sites, the Custodial Trust helped secure final development permits and sold an 18-acre parcel to Minnesota-based Dayton Hudson Corporation for a record sale price of $640,000 per acre of buildable land. Dayton Hudson Corporation then constructed a 200,000 square-foot retail center anchored by Target Stores at the site.

The Custodial Trust subsequently sold two additional parcels for development of a 900,000 square-foot, high-end office park and hotel complex. Today, this brick and glass corporate campus is home to Raytheon Corporation, Marriott Hotels and several other companies. The remediated wetlands and Hide Piles, which look like grass-covered hills, provide scenic open space.

By the mid-1990s few superfund properties had ever been sold. The significant risk of being named a potentially responsible party under the "strict, joint and several" liability scheme of the US hazardous waste clean-up laws has had a chilling effect on the sale of most superfund property. Typically, the multi-million dollar, ever-escalating costs to clean-up a superfund sites far exceed the value that can be derived from the real estate. At Industri-plex the USEPA's original estimate for the clean-up was $13 million; to date, the Remedial Trust has spent close to $90 million. Even the dramatic increase in total Custodial Trust property value—from an initial assessed value of $1,000,000 "as if clean" to actual sales of almost $17 million "as is and with all faults"—does not approach the cost of the remedy.

The specter of such liability led all Custodial Trust buyers to insist upon (as a condition precedent to closing) receipt of a Prospective Purchaser Agreement (PPA) from USEPA for superfund liability protection. Under the PPAs, USEPA granted a "covenant not to sue" for existing contamination in exchange for the buyers' cooperation on the remedy, grant of access, and inauguration of and compliance with Institutional Controls. The liability relief afforded under the USEPA PPA program was critical to ensuring that innocent purchasers of this superfund property, as well as their tenants, lenders and successors-in-title, could avoid becoming responsible parties under Superfund.

2.3 Redevelopment results

The economic rewards and other public benefits resulting from the Industri-plex redevelopment far exceeded most stakeholder expectations. To date, the Custodial Trust has distributed sale proceeds of almost $17 million to its three designated beneficiaries: the City of Woburn; the Remedial Trust; and the USEPA. Table 1 (developed by USEPA) summarizes the jobs and state and local taxes that have been created by this project. A study, commissioned by the City of Woburn, predicts that redevelopment of the Site and the resulting development of surrounding properties will create as many as 12,000 new jobs by 2010.

Beyond the positive economic impacts, this redevelopment has created many other benefits for the community and the environment. The Anderson RTC public transit services have helped to bring the state of Massachusetts into compliance with the US Clean Air Act by eliminating the air emissions from up to 2,500 vehicles commuting to and from Boston. Additionally, by constructing the intensive RTC Alternative Cover (in lieu of the Basic Soil Cover), the Transportation Agencies' remedy is more permanent and protective.

Most important, however, is that Woburn, Massachusetts has been re-born. Once characterized by former six-term Mayor John Rabbitt as, "the albatross of the Woburn," Industri-plex is now the economic center of this community. According John Rabbitt's successor, Mayor Robert Dever, "The transformation of this 245-acre site has restored Woburn's pride, hope and economic future."

Industri-plex is not about forgetting the past; rather, as evidenced by the Anderson RTC, it has helped write the next, good-news chapter in Woburn's history.

Table 1: Economic results of redevelopment (USEPA) [2].

Short Term	▪ Average of more than 700 short-term jobs per year for two years of redevelopment, with estimated total annual income of $23.6 million
Long Term	▪ As many as 4,300 permanent jobs created by redevelopment ▪ Approximately $147 million in annual income associated with permanent jobs, expected to result in almost $118 million in personal spending
Public Revenue	▪ More than $13 million in State income and sales tax associated with spending ▪ More than $1.5 million in State retail sales taxes annually ▪ More than $4 million in local annual real estate taxes
Property Value	▪ Roughly $4.6 million potential increase in residential property values within two miles of site

3 Lessons learned

3.1 Community-based redevelopment

Communities have the most to lose or gain from the remediation and redevelopment of hazardous waste sites within their borders. In Woburn, Massachusetts, where the burden of contamination, stigma and mistrust seemed insurmountable, the stakes were huge. Nevertheless, availing themselves of the technical expertise and other resources provided by the Custodial Trust, the community was empowered by a clear understanding of their options. Given the choice, community and civic leaders elected to participate and lead. The Custodial Trust's alliance with the community was critical to its efforts to secure state and federal support and funding for a major new infrastructure projects.

A study of superfund land-use and remedy-selection by the Washington, DC-based not-for-profit research organization, Resources for the Future (RFF), analyzed the Industri-plex Site prior to actual Site redevelopment. It observed that "The Custodial Trust…has actively courted the community, involving legislative representatives at the local, state, and federal levels, cultivating contacts with the media, hosting or participating in local events…and attending public meetings on Industri-plex and often playing a major role in presenting and explaining developments at Industri-plex to the public. By most accounts, it has quite effectively mobilized Woburn citizens interested in development at the site" [3]. In another report, RFF states that, "The Trust has taken the lead at Industri-plex on redevelopment and promoted the most successful recent public involvement. It has made the [Anderson] RTC possible, with its political activities and garnering of public support, including its successful work with the transportation agencies to gain approval in the state legislature for $20 million in bond bills to build the RTC and the new I-93 interchange" [4].

From the Custodial Trust's perspective, its alliance with the City of Woburn and the community's ability and willingness to make the case for these vital improvements were critical to the project's success.

3.2 Creating value for all Custodial Trust beneficiaries

The Custodial Trust pursued a host of strategies that resulted in the creation of significant, previously non-existent value for all stakeholders, some of which are highlighted below:

3.2.1 Integrating re-use in the remedy

Integrating re-use into the final Site remedy, best exemplified by the Anderson RTC, not only created new stakeholder value, it also helped ensure that that the redevelopment would be safe. The Remedial Trust realized significant remediation cost savings due to the shifting of costs from the remedy to the redevelopment. Additionally, the State Transportation Agencies have relieved the PRPs of potential future liability for the long-term stewardship of some of the Site's most contaminated areas. Finally, the Anderson RTC helped create significant real estate value by answering prospective purchaser concerns about the efficacy of the remedy as well as the stigma of the property.

3.2.2 Managing risk and minimizing liability

For prospective purchasers, tenants and lenders of Custodial Trust property, the PPA was an essential tool for managing their risk and avoiding superfund liability. The PRPs, concerned that reuse of the Site could give rise to possible future superfund liability, needed assurance that the purchasers had the resources and capabilities to comply with Institutional Controls. Accordingly, the Custodial Trust sought sophisticated buyers with financial depth and wherewithal to address the PRPs' concerns—buyers such as Raytheon Corporation, Target Stores, Marriott Hotels, even the Transportation Agencies.

3.2.3 Multi-stakeholder alignment and public-private partnerships

The structure of the Custodial Trust helped to align the diverse public and private interests. Vested with the rights and obligations of a private landowner, the Trust has direct fiduciary responsibility to the three major stakeholders to any superfund site—the community, the PRPs and the government regulators. Pursuant to its mandate, the Custodial Trust crafted solutions that balanced and satisfied the diverse goals of its three beneficiaries: (i) Woburn's desire to improve public transportation, create new jobs and tax revenues and overcome the stigma of superfund; (ii) the Remedial Trust's objective of maximizing sale proceeds to help offset the clean-up costs while minimizing future liability; and (iii) the USEPA goal to remediate the Site, and then, to the extent feasible, allow reuse in a manner protective of public health and the environment.

Once it had arrived at a reuse plan that was supported by its three beneficiaries, the Custodial Trust was positioned to seek and negotiate multiple public-private agreements that resulted in the ambitious infrastructure improvements at the Site. As explained by USEPA, "The Custodial Trust has worked extensively with federal, state, and local officials and the neighboring communities to promote the redevelopment of the Industri-Plex site and its surrounding areas. At the heart of this redevelopment project was the public and private partnership that generated broad support for improvements at the site" [2]. The public-private partnerships that resulted in ~$50 million in major publicly funded infrastructure were critical to creating significant real estate value and overcoming market fears about the stigma of the property.

3.3 The role of an independent third party – the Custodial Trust

According to Resources for the Future (RFF), "Probably the most important feature contributing to the successful reuse [of Industri-plex] has been the custodial trust. This legal entity…has been an effective and innovative force that has brought to the cleanup process the development interests of local government, state agencies, and the private sector" [5].

As custodian and landowner of more than half of the 245-acre Site, the Custodial Trust has effectively functioned as an independent advocate for Site redevelopment, able to work outside the traditional superfund enforcement process. RFF notes that, "In creating the custodial trust, the consent decree effectively severed liability from the trust's redevelopment activities. The removal of liability has allowed the trust the opportunity to create private/public partnerships to attract large-scale retail outlets to the site…and to encourage the public and their elected officials to discuss reuse options. Large-scale development, we now know, can take place at a heavily contaminated NPL [superfund] site" [5].

Severing the Custodial Trust from superfund liability has indeed been a crucial ingredient to the success of Industri-plex. But more important was the Trust's ability to forge innovative agreements and initiatives that aligned the interests of many diverse stakeholders. It is worth noting that the Custodial Trust is strictly prohibited from engaging in commercial activities. Also, rather than taking an equity interest in the property, the Trust is compensated by management fee, thus avoiding any real or perceived conflict in the potentially competing goals of Site remediation and redevelopment.

4 Conclusions

Large, multi-faceted, multi-party sites with complicated contamination issues are often referred to as "upside down" properties (i.e., properties where the value of the real estate is significantly less than the cost of the clean-up). These sites require value creation strategies—such as integrating future use into the remedy, managing risk, minimizing liability and aligning key stakeholder interests—to overcome the barriers to reuse. The creation of an independent, third-party entity, such as the Industri-plex Custodial Trust, is also a widely applicable and useful mechanism for redeveloping large sites with complex contamination issues, property gridlock, or diverse and competing interests that require multi-stakeholder alignment and community-based solutions.

In a letter in support of the Industri-plex nomination for the prestigious National Phoenix Award, the Reverend Bruce Young, former minister for the Woburn Trinity Episcopal Church, wrote, "The Industri-plex Site will stand as a beacon of hope and offer a blueprint of management excellence to those who look at brownfields and ask, 'How can we ever remediate this site? It is so large and complex.' The answer—the model is found at Industri-plex."

References

[1] Hemond, Harold F. "Movement and Distribution of Arsenic in the Aberjona Watershed," *Environmental Health Perspectives*, pp. 36, Volume 103, Supplement 1, February 1995.

[2] Harr, Jonathan, *A Civil Action*, Random House Books, New York, 1997.

[3] Industri-plex Site Case Study, Superfund Redevelopment Initiative (SRI), U.S. Environmental Protection Agency (USEPA), OSWER 9378.0-19FS, Office of Solid Waste and Emergency Response, Online. www.epa.gov/superfund/programs/recycle/success/casestud/iplexcsi.htm

[4] Wernstedt, K., Hersh, R., & Probst, K, "Basing Superfund Cleanups on Future Land Uses: Promising Remedy or Dubious Nostrum?" Discussion Paper 98-03, Resources for the Future, Washington DC, pp. 19, Oct. 1997.

[5] Wernstedt, K. & Probst, K, "Land Use and Remedy Selection: Experience from the Field—The Industri-plex Site," Discussion Paper 97-27, Resources for the Future, Washington, DC, July 1997.

[6] Hersh, R., Probst, K, Wernstedt, K., & Mazurek, Jan, "Linking Land Use and Superfund Cleanups: Uncharted Territory," Resources for the Future, pp. 43, June 1997, Internet Edition Online. www.rff.org.

Section 9
Monitoring of contaminated sites

Remote sensing detection techniques for brownfield identification and monitoring by GIS tools

C. Atturo[1], C. Cianfrone[2], V. Ferrara[3], L. Fiumi[1], G. Fontinovo[1] & C. M. Ottavi[3]
[1]*LARA, IIA CNR, Area della Ricerca di Roma 2 Tor Vergata, Italy*
[2]*Faculty of Engineering, University of Rome "La Sapienza", Italy*
[3]*Department of Electronic Engineering, University of Rome "La Sapienza", Italy*

Abstract

This paper examines remote sensing detection techniques for data integration in GIS to be employed in brownfield monitoring. Brownfields are frequently associated with urban wastes, mostly from building demolition. The hazard increases when dangerous materials are used. In order to allow us to perform a detailed classification of the land and to identify a wide range of the brownfields, the Laboratory of *CNR* (LARA) detects environmental data by means of the airborne MIVIS (*Multispectral Infrared and Visible Imaging Spectrometer*) sensor. Moreover, GIS tools correlate the relevant data in order to evaluate the extension and derivative parameters of the same brownfields. These parameters are then used to elaborate indicators that can be used within the framework of a sustainable planning and development.
Keywords: monitoring of contaminated sites, multispectral sensors, MIVIS, GIS, multimedia modelling.

1 Introduction to brownfield identification

Once thriving areas of economic activity, brownfields are today generally abandoned/underused urban and sub-urban zones. Their expansion can be risky because it frequently involves some sort of environmental contamination, depending on the wear state of the building taken into consideration. They

WIT Transactions on Ecology and the Environment, Vol 94, © 2006 WIT Press
www.witpress.com, ISSN 1743-3541 (on-line)
doi:10.2495/BF060231

unavoidably create an empty architectural space with environmental and urban concerns. Brownfield sites, especially if urbanized (that is, with the utility networks still in place), and if well-linked, can be restored to productive use both from an environmental and economic viewpoint, thanks to the lower soil use .

The remote sensing classification techniques allow us to identify areas particularly at risk. The process of mapping of the roofing material and the detection of hazardous substances both in rural and in urban areas, represent the first and most important step to assess the risk level related to the study area under examination. As a matter of fact, these detected substances, should be at first considered as potentially hazardous only. The wear state, the maintenance and the repair of a building structure could reduce the related risk level. The interpretation of remotely sensed data must always be associated to *in situ* measurements because they are often decisive to distinguish among different kinds of situation and to provide a more accurate risk level assessment. These techniques, however, can considerably reduce the human resource employment and the related costs because they require a far less extensive ground campaign care must be taken however to validate the resulting risk level with reliable data.

When hazardous substances and materials are detected, a cross-verification technique should follow, i.e. a correlation among different territorial parameters or among indicators. It is evident that a substance becomes more hazardous when it is closer to other elements or buildings emphasizing their diffusive and explosive characteristics. Soluble substances detected close to springs, streams or simple canals, because of torrential phenomena, cannot be treated as those away from any element emphasizing their diffusive characteristics.

The main characterization require for brownfields often concerns the abandoned areas and their wear state. The presence of hazardous substances in the same area makes the situation even worse. A specific indicator should be provided in order to suggest, according to the risk level, the restoration monitoring as a priority. Within the framework of the urban areas, the Public Authorities entrusted of environmental monitoring can easily implement the recognition as well as the identification of wear state conditions. This thanks to the easy access to the area and to the protest of those who live close to brownfields. In these areas the automated or semi-automated techniques can, if this is the case, help identifying areas at risk. Whereas the same techniques can prove to be significantly more favourable for those less populated (or harder to access areas with recent wear state situation (often in rapid evolution).

The hyperspectral remote sensing, already intended for the territory monitoring and the relative qualification, provides essential data for further territorial classification. Basically all the pixels associated to the classified material are tagged with a brownfield label The pure recognition would make desirable a series of consecutive and derivative steps: extension evaluation, possible risk assessment according to the above-mentioned specifications, nearness evaluations and so on. All these steps have to be implemented by means of an appropriate GIS. The above automated or semi-automated techniques are implemented by the post-processing image coming from airborne remote sensing (in this case, based on hyperspectral data). These images are correlated, if needed, to others originating from a different source, geo-referred as the first ones and so overlaid over the same area even if having a different resolution. This could provide a more detailed recognition and a lower uncertainty of the geometric features.

2 Brownfields: hyperspectral classification techniques

To support the research activity, the Italian National Research Council has established the LARA Project (Airborne Laboratory for Environmental Research) with the aim to remotely collect data by means of an airborne platform system called MIVIS. It is a modular scanning system that operates with a high spectral and geometric resolution. It is composed of 4 spectrometers able to simultaneously collect samples in 102 spectral bands within 0.43 μm and 12.7 μm. This allows the study of both the chemical and physical parameters of the Earth's surface. The MIDAS (*Multispectral Interactive Data Processing and Analysis System*) station converts data at ground and it is capable of performing the radiometric correction. The hyperspectral data processing and classification are aimed at characterizing several ground features of general interest such as building structures, road paving, crops, bare soils, vegetation, wastes and so on.

Table 1: MIVIS sensor characteristics.

Spectrometers	Spectrum	Bands	Spectral interval (μm)	Bandwidth (μm)
I	Visible	20	0.43-0.83	0.02
II	Near Infrared	8	1.15-1.55	0.06
III	Mean Infrared	64	2.0-2.5	0.009
IV	Thermal Infrared	10	8.2-12.7	0.34-0.54

The discrimination of waters and, as earlier mentioned, the identification of the characteristics related to abandoned or otherways investigated areas is based on the *spectral signature* concept: all the natural and artificial objects have different emission properties within the various optical bands used; they also exhibit and different values of *backscattering* from different radio waves. As a consequence, modern remote sensing systems are implemented taking into account the possibility of varying the observation parameters: the object is observed through different sensors varying the frequency, polarization and the incidence angle of the scenes. An appropriate integration of these hyperspectral data will follow, allowing us to extrapolate information otherwise not visible on data taken from single measures. Once acquired the raw data, a pre-processing phase is required. During this phase there is an attempt to reduce the following process of the scenes due to alignment problems of the different operating sensors; contrasts are altered or emphasized as needed, and the effects of high frequency attenuation are compensated. In addition, operations such as geometrical and co-ordinate transformations can be performed. Finally, the correction of atmospheric effects together with *IRC* (Internal Radiometric Calibration) are implemented: the retro-diffused energy coming from the atmosphere towards the sensor can be considerably important. The radiometrically corrected is then processed using the *supervised* classification technique called *Spectral Angle Mapper* (SAM). This method enables us to quickly map the similarities existing between image spectra and reference spectra [2]. The reference spectra can be determined either in laboratory or in field; otherwise they can be extracted from images, (see Figure 1). The algorithm gives the spectral similarity between the two spectra in terms of the "angle" which they form, and, therefore, the spectra are considered as vectors in a space whose dimension is equal to the *nb* (number of bands). The similarity between an unknown spectrum t and a reference spectrum r comes from the following equation:

$$\alpha = \cos^{-1}\left(\frac{\vec{t} \bullet \vec{r}}{\left\|\vec{t}\right\| \bullet \left\|\vec{r}\right\|}\right)$$

that can also be expressed as follows:

$$\alpha = \cos^{-1}\left(\frac{\sum_{i=1}^{nb} t_i r_i}{\left(\sum_{i=1}^{nb} t_i^2\right)^{1/2}\left(\sum_{i=1}^{nb} r_i^2\right)^{1/2}}\right)$$

where α is the angle between vectors and *nb* is the number of bands in the image.

For each reference spectrum chosen in the analysis of hyperspectral images, the angle α is determined for every element of the image (pixel). This value in radiants is assigned to the corresponding spectrum in the output SAM image, one for each reference spectrum. The maps of the produced spectral angle show a new cube of data providing a band number equals to that of the reference spectra used for the mapping. The SAM algorithm, implemented in ENVI software, requires as an input a number of training areas or reference spectra resulting from specific "Regions Of Interest" (ROIs) or spectral databank. As far as this study is concerned, the input spectra have been extracted from ROIs accurately identified in the view, integrated with a series of spot observations of the study areas and visual analysis of additive syntheses in RGB (*Red*, *Green*, *Blue*), (see Figure 2). When building a GIS based on MIVIS data, the land scheme is implemented by adopting two different approaches: the vector and the raster data technique. In the raster model the representation of the objects is based on the co-ordinates (row and column) of the pixels of the matrix forming the image. In the vector model the representation of the objects is based on features: points, lines, segments, areas, polygons, etc. Once the image is classified in significant classes by means of *supervised and/or unsupervised* spectral methods, it becomes a true thematic map because the classification is based on characterization classes. The integration of the hyperspectral data in *GIS* (Geographic Information System) (see Figure 3), is one of the most effective applications of data integration. The morphological, chemical and physical features of the study area, can be accurately superimposed on the geographical maps to allow us to closely examine the interactions underway in the investigated or expected in the future. We would like to stress that the integration process of hyperspectral data, in the territorial information systems, has always been difficult when compared with a common remote sensing data. This depends on the nature of the employed data: in a reduced data set, there is always a loss of spectral information, relative to the original raw data.

Figure 1: Example of brownfield classification obtained by SAM method.

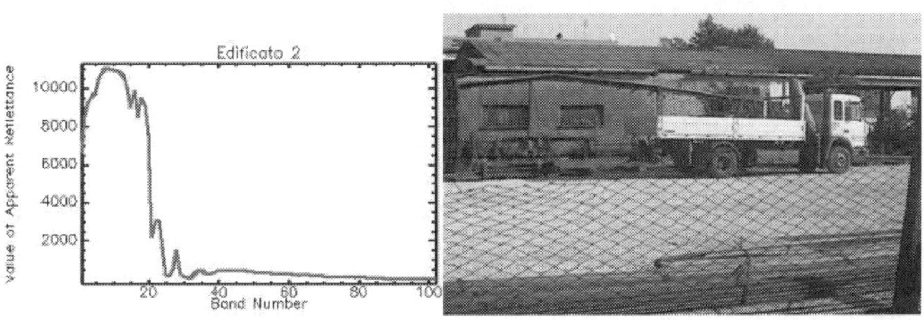

Figure 2: Building spectrum taken from MIVIS data, RGB image related to an urbanized area.

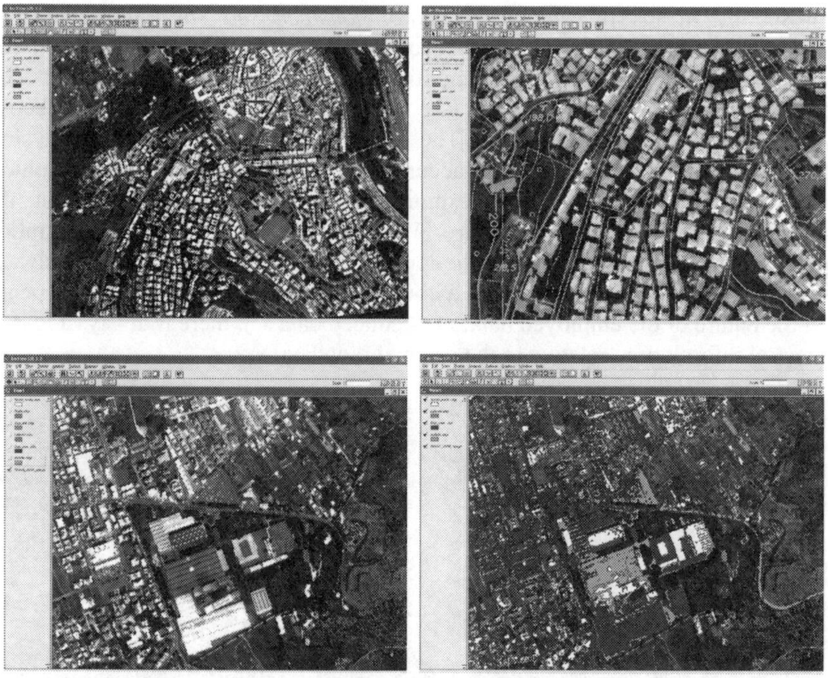

Figure 3: MIVIS data integration in GIS by adopting the raster and the vector models.

3 Identification procedures: convolution and filtering techniques

In most cases of satellite/airborne remote sensing, buildings pulled down (or abandoned) together with waste lands show irregular contours of the structure and discontinuities of the surface coverings. Detection techniques evaluate the homogeneity of lines and surfaces. In any case they will have to be optimized in terms of timing, tools and response reliability level.

Of course, we could resort to airplane campaigns intended for a software 3-dimensional reconstruction of topography, which could make easier the recognition of the wear state buildings. These campaigns are aimed at *a)* the DEM (*Digital Elevation Model*) achievement and *b)* a 3-D software model reconstruction of highly urbanized areas. In the former case, remote sensing systems are normally used. The new algorithmic techniques reduce the production costs of these images, obtaining, for instance by means of AsterDTM [6], a 15-m pixel resolution, mean errors on the 5 m altitude and final costs related to image within the range of some hundred euros. In the latter, there is the need for a more accurate elevation range, so the airplane campaigns are essentially intended for urbanized areas. About brownfields, alternative techniques are adopted. As previously mentioned, technicians use hyperspectral data and higher spatial resolution images, if necessary, in the Visible. An image detected in the Visible provides as database some information sensitive to geometric features, angles of the various facets whose the surface coverings is eventually composed. Continuous lines and surface homogeneity are the parameters to distinguish between a good building conservation and a highly abandoned area.

The identified procedure has used two parallel levels: the first one emphasizes the building contours, whereas the second one evaluates the indicators of the surface homogeneity level.

As far as the contours are concerned, we adopted classical filtering techniques using *kernel*, for example those introduced by Sobel and Prewitt (together with multi-stage process and related procedures introduced by Canny [3] and Parker [9]). *Kernels* operate on a pixel grid, representative of a coded image in different grey levels, and they are generally two matrices: one matrix is simply the other rotated by 90°. When each *kernel* is applied, it emphasizes the different grey levels towards the two main directions: from left/right to right/left (Sx) and from down/up to up/down (Sy). In the case of Sobel *kernel,* matrices are as follows:

$$Sx = \begin{bmatrix} -1 & 0 & +1 \\ -2 & 0 & +2 \\ -1 & 0 & +1 \end{bmatrix} \qquad Sy = \begin{bmatrix} +1 & +2 & +1 \\ 0 & 0 & 0 \\ -1 & -2 & +1 \end{bmatrix} \tag{1}$$

The convolution operation between the matrix of the image I and the Sobel *kernel* provides the values related to the two matrices Gx and Gy, eqns (2) [7]. These can be normally combined together to find the absolute magnitude of the gradient, eqn (3), or in the approximated form by eqn (4) which detects the researched edge.

$$Gx_{m,n} = \sum_{k=1}^{3}\sum_{p=1}^{3} I_{m+k-2,n+p-2} Sx_{k,p} \quad ; \; Gy_{m,n} = \sum_{k=1}^{3}\sum_{p=1}^{3} I_{m+k-2,n+p-2} Sy_{k,p} \tag{2}$$

$$|G| = \sqrt{Gx^2 + Gy^2} \tag{3}$$

$$|G| = |Gx| + |Gy| \tag{4}$$

Before applying the convolution technique and in order to obtain results significantly useful for identifying the same edges and defining the borders, an intensity thresholding filtering is implemented. These filters can employ either fixed or adaptive threshold value. The procedure starts from the hyperspectral classification: the related maps allow one to identify the involved areas showing covering materials with possible potentially hazardous substances. These areas are geo-referred and virtually superimposed on the map referring area in the Visible (a grey-level image with a higher resolution). The different resolution between the two maps will determine a non-biunivocal correspondence: one pixel (hyperspectral map)→ more pixel (grey-level map). In every selected area the statistics of the grey levels is evaluated. The relative pixels are aggregated into groups according to the intensity level they match. In the case of a good building conservation, histogram intensity levels consist of evident few peaks as shown in the example of Figure 4. Each group of the pixel intensity distribution produces a threshold by means of the mean value determination and the identification of a standard deviation. This filtering process is an evolution of the filtering method based on adaptive threshold value and it is mostly intended to achieve the main objective of this research activity.

The convolution operation with selected *kernels* outlines the boundaries among pixel with a different intensity. If we take into consideration buildings in good conditions, these boundaries should produce closed curves. But normally they could yield discontinuous lines. The number of interruptions, i.e. the absence of a regular sequence of contour, identifies an indicator of the building wear state or the lack of building walls. The above-mentioned discontinuities could be reduced by the thematic hyperspectral maps: equivalent responses in the Visible are discriminated from the different recorded qualification.

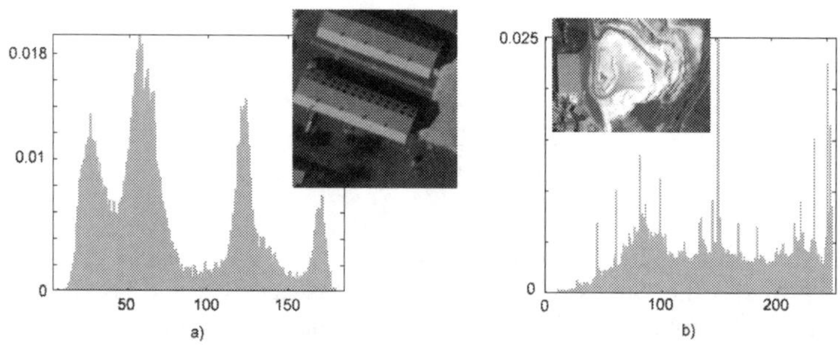

Figure 4: Two examples of histogram for typical structures: a) buildings that are in good state of conservation; b) waste lands.

Building roof mappings, for instance, differ from asphalt and grassy surfaces but, due to the colour and brightness effect, they show the same grey-level intensity. The vectorialization of the boundary sequence offers computational benefits in the following operations to verify the area to which the specific pixels belong. This transformation, however, could be burdensome and the previous operations could be replaced by *raster*-based equivalence classes. Together with this edge detector technique is assessed the non homogeneous levels within the study area: the non uniformity of the spatial distribution relates to covering materials such as cement, coating materials, etc. This is clearly the most cost-effective procedure, if we consider the rubbish open-air depot monitoring. The non uniformity analysis can start directly from the previously mentioned thresholded filtering data or it can be evaluated by the intensity hystogram of the original image in the Visible. In particular, we get rid of operators designed to determine the different frequencies within the study area.

The application of a FFT (*Fast Fourier Transformer*) provides the expected output, but, unfortunately, this method transforms the domain of the representation losing the information on the original spatial position [8]. A Wavelet transform [4, 5] allows to obtain a response which can be also located and generally applied to a more extended area. Usually, wavelet is associated to the scaling function. The last one is evaluated by convolving low-pass quadrature mirror filters, and when used in wavelet decomposition, they realize the *approximation*. Complementary, wavelet is valued by high-pass filter which produces the *details*, i.e. the discrepancy between two images generated by consecutive application of an orthogonal wavelet analysis. In the case of two-dimensional context, wavelets are defined as tensor products of scaling function and three wavelets, in three orientations, are generated: horizontal, vertical and diagonal.

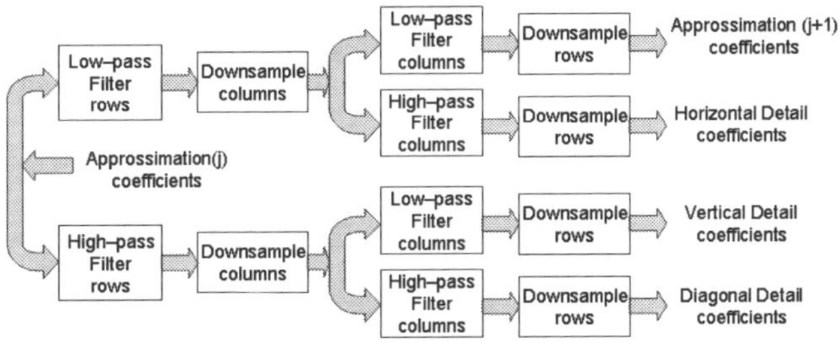

Figure 5: Wavelet: schematic algorithm of decomposition step.

a) b) c) d)

Figure 6: Sequence of the building edge construction : a) Grey-level image of a block; b) after the thresholded filtering technique; c) application of the Sobel kernel convolution; d) reconstruction of discontinuous outlines and possible elimination of internal sides.

Figure 5 shows an algorithm of each decomposition step and the related dyadic decimations. Consequently, the following outputs are coarser images. Detail coefficients are localized and could help to evaluate the above-mentioned operators, designed to determine the frequencies. Figure 6 shows the sequence of the building edge construction. To the grey-level image is applied a thresholded filtering technique. Thresholds have been chosen on the local base and according to the considered materials only. If we consider the example, we can note that the match "covering material and brightness intensity in the Visible" has identified three thresholds which generate Figure 6 b). The convolution operation with Sobel *kernel* produces contour sequences which are partly interrupted and unfortunately they identify objects not belonging to the study area. The relationship with the qualification data allow to isolate the considered edges only in order to reconstruct the discontinuous outlines and, if necessary, to eliminate the internal sides.

4 Conclusions

The results obtained within the framework of this study make foresee excellent potentialities of MIVIS hyperspectral data and GIS to monitor brownfield areas. The use of MIVIS data proved to be particularly interesting for identifying brownfield nature by means of classification techniques. Furthermore, when hyperspectral data processed in GIS are integrated, a huge potentiality of these systems is shown in integrating and correlating data from different sources to be used within the framework of a sustainable planning and development.

Acknowledgements

The authors wish to thank Mrs Alessia Turini for the English translation of the text.

References

[1] Ben-Dor, E., Kruse, F. A., Lefkoff A. B. & Banin A., Comparison of three calibration techniques for utilization of GER 63 channel aircraft scanner data of Makhtesh Raamon Negev, Israel, *PE&RS*, **60(11)**, pp.1339-1354, 1994.

[2] Boardman, J.W. & Kruse, F.A., Automated spectral analysis: a geological example using AVIRIS data, North Grapevine Mountains, Nevada, *Proc. of the Tenth Thematic Conference on Geologic Remote Sensing, San Antonio Texas USA*, I:407-418, 1994.

[3] Canny, J., A computational Approach to Edge Detection, *IEEE Transaction on Pattern Analysis and Machine Intelligence*, **8(6)**, pp. 679-698, 1986.

[4] Chui, C.K., *Wavelets: a tutorial in theory and applications,* Academic Press., 1992.

[5] Daubechies, I., Cohen, A., Jawerth, B. & Vial, P., Multiresolution analysis, wavelets and fast wavelet transform on an interval, *CRAS*, Paris, Ser. A, t. 316, pp. 417-421, 1993.

[6] Gatti, M., Stereoscopia ottica ASTER ed interferometria differenziale SAR a confronto, *GEOmedia*, GEO4all, **1**, pp. 6-13, 2005.

[7] Gonzales, R. & Woods, R., *Digital Image Processing,* Addison Wesley Publishing Company, 1992.

[8] Haykin, S., *Adaptive Filter Theory*, Prentice-Hall, Inc., 1996.

[9] Parker, J. R., *Algorithms for Image Processing and Computer Vision*, John Wiley & Sons Inc., New York, pp. 23-29, 1997.

Evaluation for estimating partition coefficients of alcohol tracer partitioning between benzene, toluene, ethylbenzene, and xylenes (BTEX) compounds and water

S. Rhee, S. Cho & J. Park
School of Civil, Urban and Geosystem Engineering, Seoul National University, Korea

Abstract

The partitioning tracer method has been studied in laboratory research and field investigations as an alternative method for characterizing aquifers contaminated by nonaqueous phase liquids (NAPLs). Accurate partition coefficients of tracers partitioning between NAPL and water are needed to improve the reliability of the partitioning tracer method. In this research, partition coefficients of alcohol tracers partitioning between benzene, toluene, ethylbenzene, and xylenes (BTEX) compounds and water are estimated using the approach of equivalent alkane carbon number (EACN). General agreement was observed between the measured and estimated partition coefficients of alcohol tracer between BTEX compounds and water. Based on these results we can verify that the EACN approach is suitable for estimating the partition coefficient of alcohol tracers between BTEX compounds and water.
Keywords: partitioning tracer method, alcohol tracers, partition coefficient, BTEX compounds.

1 Introduction

The contamination of soil and groundwater by nonaqueous phase liquids (NAPLs) has become an issue of great interest. In particular, petroleum, such as gasoline, has been used in large quantities in various industry field and its spills or leaks into aquifers cause environmental problems. The major components of petroleum are benzene, toluene, ethylbenzene, and xylenes (BTEX) and they

www.witpress.com, ISSN 1743-3541 (on-line)
doi:10.2495/BF060241

have known to be carcinogenic organic contaminants [1]. If petroleum leaks into the ground from a storage tank it is necessary to monitor the distribution and quantity of petroleum and also to cleanup contaminants immediately [2].

However, traditional site characterization methods, such as analysis of soil gas, core sampling, and monitor-well sampling, are often not capable of providing sufficient data required for effective remediation system design. A major weakness of these methods is that they provide data at discrete points. As a result there is a probability of under-estimation of NAPL's residual saturation without a cost-prohibitive amount of sampling [3].

The partitioning tracer method has been developed and demonstrated in the field during the past few years as an alternative method for site characterization. The major advantage of the partitioning tracer method is to provide a continuous measurement of the target area and characterize a relatively large and reliable detection volume in comparison with traditional methods [4, 5]. The partitioning tracer method consists of the simultaneous injection of several tracers, which have different partition coefficients with respect to NAPL at one or more injection wells and subsequent measurement of tracer concentrations at one or more extraction or monitoring wells. Injected tracers are retarded in proportion to its partition coefficient between NAPL and water due to their chromatographic separation in aquifers. This separation of the partitioning tracers indicates the presence of NAPL in the targeted zone and is used to determine the volume and distribution of NAPL present [4].

The accuracy of a partitioning tracer method mainly depends on the selection of suitable tracers and the accurate measurement of their partition coefficients between NAPL and water. Dwarakanath and Pope [5] suggest that the error in NAPL volume estimates based on partitioning tracer methods is directly proportional to the error in the partitioning coefficients. Therefore, the accurate experimental measurements of partition coefficients are important to minimize the errors in the partitioning tracer method. However, it requires time-consuming experiments to obtain the partition coefficient of each proposed tracer. Moreover, if samples of NAPL are not available, it is difficult to measure the partition coefficient directly.

Dwarakanath and Pope [5] proposed a new approach for estimating partition coefficients of alcohol tracers between NAPL and water using the concept of equivalent alkane carbon number (EACN), which is simply an equivalent of the alkane carbon numbers for more complex organic molecules (e.g., 6 for hexanol), and suggested that it is very effective and useful in estimating the partition coefficient of alcohol tracers between water and NAPLs. They also suggested the bilinear equation by multiple regression analysis on the basis of the results that the correlation between the logarithm of the partitioning coefficient of various alcohols and EACN of NAPLs and alcohol tracers is a linear relationship. Wu and Sabatini [6] have extended this approach and focused on more hydrophobic oils (high EACN LNAPLs) to find the hydrophobicity of NAPLs. These studies mainly focus on dense nonaqueous phase liquids (DNAPLs) and high EACN number of LNAPLs.

In order to estimate the partition coefficient of alcohol tracers between petroleum and water by using EACN approach, BTEX are selected to represent gasoline as petroleum-contaminants and the EACN approach is applied for estimating the partition coefficients of several alcohol tracers for various BTEX mixtures. To confirm the accuracy of the EACN approach for petroleum, and in particular for gasoline, the partition coefficients from experimental measurements are compared with the estimated values from the EACN approach.

2 Materials and methods

2.1 Materials

The properties of benzene, toluene, ethylbenzene, and xylenes and alcohol tracers used in this study are summarized in table 1. BTEX mixtures are made with certain mixing ratios given in table 2. Mixtures A to D are blended for representing the mixtures with a certain mole ratio and Mixture E is blended with the special mixing ratio, which represent the composition ratio of BTEX found in soils contaminated by petroleum [7].

Table 1: Properties of BTEX and tracers used in this study.

Chemicals	Molecular formula	Molecular weight (g/mol)	Density (g/mL)	Solubility (mg/L)	Manufactured company	Purity (%)
Benzene	C_6-H_6	78.11	0.88	1780	JUNSEI	99.5
Toluene	C_6-H_5-C-H_3	92.14	0.87	500	Aldrich	99.5
Ethylbenzene	C-H_3-C-H_2-C_6-H_5	106.17	0.86	140	Aldrich	98.0
Xylene	C_6-H_4-(C-H_3)$_2$	106.17	0.87	170	Aldrich	83.0
Methanol	C-H_3-O-H	32.04	0.79	-	Duksan	99.8
1-Butanol	C_4-H_9-O-H	74.12	0.81	-	Aldrich	99.9
1-Pentanol	C_4-H_9-C-H_2-O-H	88.15	0.81	-	Aldrich	99+
Hexanol	C_6-H_{14}-O	102.18	0.81	-	Aldrich	99+
2-Ethyl-1-Butanol	(C_2-H_5)$_2$-C-H-C-H_2-O-H	102.18	0.83	-	Aldrich	98.0
4-Methyl-2-Pentanol	(C-H_3)-C-H-C-H_2-C-H-(O-H)-C-H_3	102.18	0.81	-	Aldrich	99.0

Table 2: Mixing ratios for BTEX compounds.

BTEX compounds	Mixing ratios							
	Mole ratio				Weight ratio			
	Benzene	Toluene	Ethyl -benzene	Xylene	Benzene	Toluene	Ethyl -benzene	Xylene
Mixture A	1	1	1	1	1	1.18	1.36	1.36
Mixture B	2	1	1	1	1	0.59	0.68	0.68
Mixture C	1	2	1	1	1	2.35	1.36	1.36
Mixture D	1	1	2	1	1	1.18	2.72	1.36
Mixture E	1	13.18	3.96	17.01	1	15.55	5.39	23.12

2.2 Experimental method

The partition coefficient was measured by using six tracers and nine BTEX compounds, which were benzene, toluene, ethylbenzene, xylenes, and Mixture A to E. The alcohol aqueous concentrations were varied between 100 and 600 mg/L (100, 200, 400, 600 mg/L). Equal volumes of BTEX compounds and water (10 mL each) were placed in a 20 mL vial that was crimp-capped with a Teflon septum, leaving minimum headspace. The vial was shaken on a vortex mixer and rotary shaker for each 30 seconds and 24 hours. After shaking these samples, they were allowed to stabilize for 1 hour and than centrifuged for 5 minutes to allow a complete separation of the phases. Reacted alcohol aqueous solution was translated from aqueous phase to the 2 mL vial using a pipette and were analyzed by an Agilent 6890 GC-FID (each sample was made in duplicate). Then, the partition coefficient was calculated from using a mass balance.

3 Results and discussions

The partitioning of 4-methyl-2-pentanol between BTEX compounds and water is shown in fig. 1. Linear relationships are found between the concentration of a tracer in water and that of a tracer in BTEX compounds. The other 4 tracers, which are 1-butanol, 1-pentanol, hexanol, and 2-ethyl-1-butanol, also show linear relationships between the concentration of a tracer in water and that of a tracer in BTEX compounds.

From these linear relationships the slope of the regression curve represents the partition coefficient of used alcohol tracer. All of partition coefficients for each alcohol tracer between BTEX compounds and water are listed in table 3 and the regression coefficient of results have a value of over 0.95.

The EACN values of benzene, toluene, ethylbenzene, and toluene are respectively 0, 1, 2, and 2 and those of butanol, pentanol, and hexanol are respectively 4, 5, and 6. BTEX's EACN values are from the literature [8, 9] and the tracer's EACN values are from the assumption that the EACN value of linear alcohol can be found just by its carbon number [5].

A linear relationship was observed when the logarithm of the partition coefficient of several alcohol tracers was plotted against both of the EACN values of BTEX compounds (fig. 2) and those alcohol tracers (fig. 3). These results are similar with that of Dwarakanath and Pope [5].

According to these results, the bilinear equation for correlating the logarithm of partition coefficient (K) as a function of the alcohol tracer's EACN (A) and the BTEX EACN (N) was obtained by multiple regression analysis, eqn (1).

$$\log K = -2.39829 + 0.584492 \cdot A - 0.07418 \cdot N \tag{1}$$

Based on the regression of the data, the standard deviation is 15.5% and the correlation coefficient of the regression is 0.935. This equation for BTEX compounds is analogous to that of Dwarakanath and Pope [5].

Using the experimental value of the partition coefficient for single BTEX and eqn. (1), the logarithm of the experimental partition coefficient of the BTEX mixture is compared with the logarithm of the estimated partition coefficient of the BTEX mixtures (table 4) and the accuracy of measurements is given in percent (estimated value over experimental value times 100). The overall difference and tendency between the experimental and estimated results is also plotted in fig. 4. On the 1:1 line, experimental values are equal to estimated value and most over-estimated or under-estimated values are on the 1:1.19 line and 1:0.58 line respectively. Though there is a tendency that estimated results are underestimated, most of the estimated results are similar to experimental results.

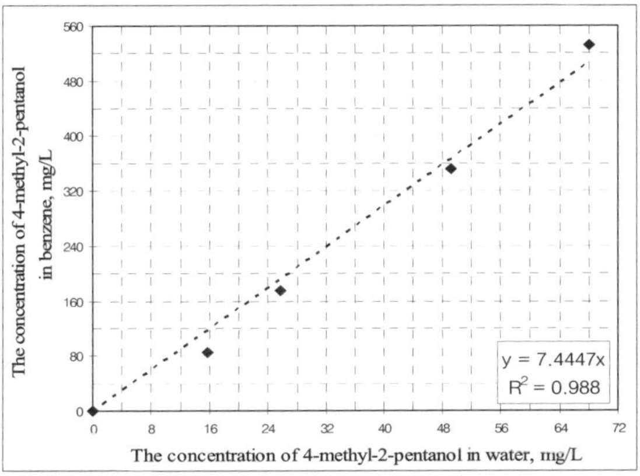

(a)

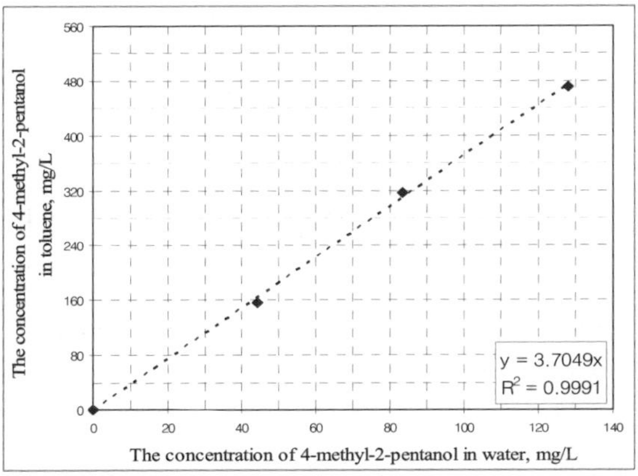

(b)

Figure 1: Partitioning of 4-methyl-2-pentanol between water and: (a) benzene, (b) toluene, (c) Mixture A and (d) Mixture B.

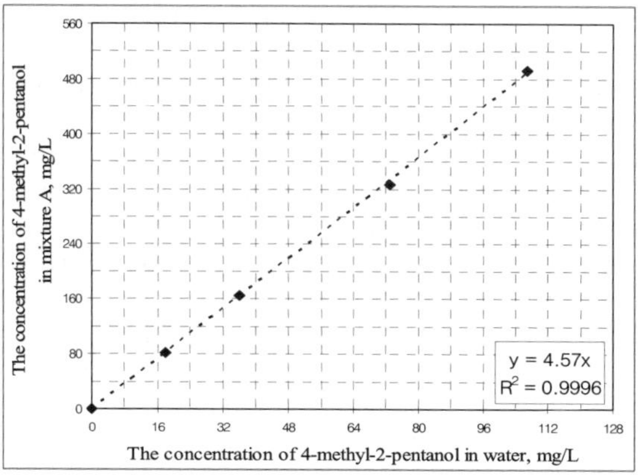

(c)

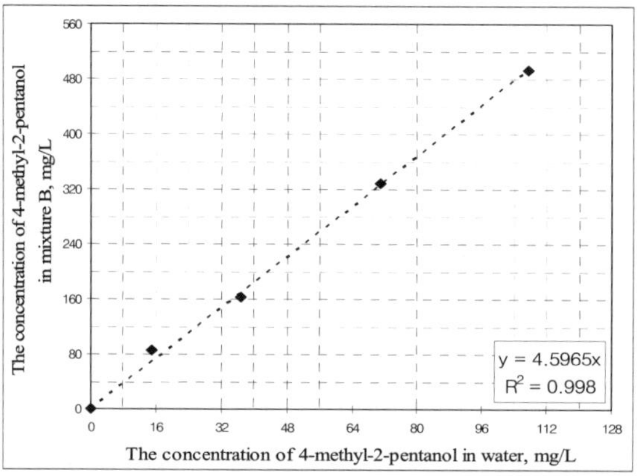

(d)

Figure 1: Continued

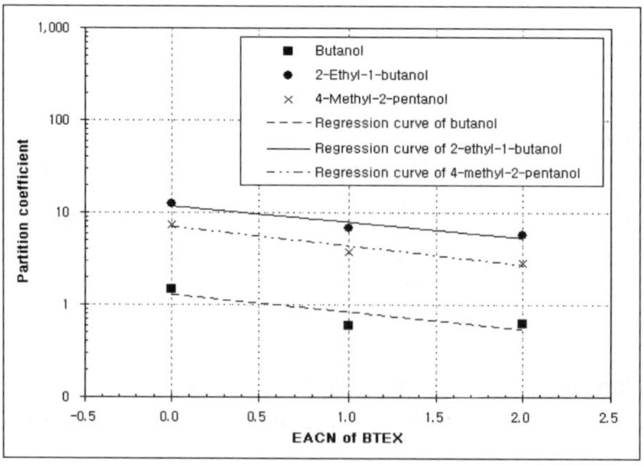

Figure 2: Partitioning coefficient as a function of BTEX EACN.

Table 3: Measured partition coefficients of tracers between BTEX compounds and water.

Tracers	Contaminants	K [a]	R^2 [b]	Contaminants	K	R^2
Methanol	Benzene	0.05	-	Mixture A	-	-
	Toluene	0.22	-	Mixture B	-	-
	Ethylbenzene	0.28	-	Mixture C	-	-
	Xylene	0.19	-	Mixture D	-	-
				Mixture E	-	-
Butanol	Benzene	1.47	0.9999	Mixture A	1.99	0.9745
	Toluene	0.60	0.9784	Mixture B	1.10	0.9955
	Ethylbenzene	0.62	0.9992	Mixture C	0.76	0.9984
	Xylene	0.88	0.9429	Mixture D	0.96	0.9719
				Mixture E	0.42	0.9983
Pentanol	Benzene	2.68	0.9793	Mixture A	2.36	0.9922
	Toluene	1.98	0.9918	Mixture B	2.49	0.9994
	Ethylbenzene	2.20	0.9915	Mixture C	2.31	0.9988
	Xylene	1.68	0.9990	Mixture D	2.37	0.9905
				Mixture E	2.88	0.9480
Hexanol	Benzene	14.00	0.9954	Mixture A	14.48	0.9679
	Toluene	8.65	0.9900	Mixture B	10.67	0.9979
	Ethylbenzene	14.51	0.9931	Mixture C	7.62	0.9952
	Xylene	12.06	0.9963	Mixture D	18.23	0.9579
				Mixture E	19.29	0.9643
2-Ethyl-1-butanol	Benzene	12.59	0.9885	Mixture A	6.97	0.9996
	Toluene	6.86	0.9903	Mixture B	6.74	0.9879
	Ethylbenzene	5.71	1.0000	Mixture C	8.07	0.9851
	Xylene	5.95	0.9995	Mixture D	6.56	0.9918
				Mixture E	5.74	0.9992
4-Methyl-2-pentanol	Benzene	7.44	0.9880	Mixture A	4.57	0.9996
	Toluene	3.70	0.9991	Mixture B	4.60	0.9980
	Ethylbenzene	2.83	0.9899	Mixture C	6.15	0.9752
	Xylene	3.48	0.9951	Mixture D	4.55	0.9883
				Mixture E	9.18	0.9488

[a] K - Partitioning coefficient
[b] R^2 - Regression coefficient

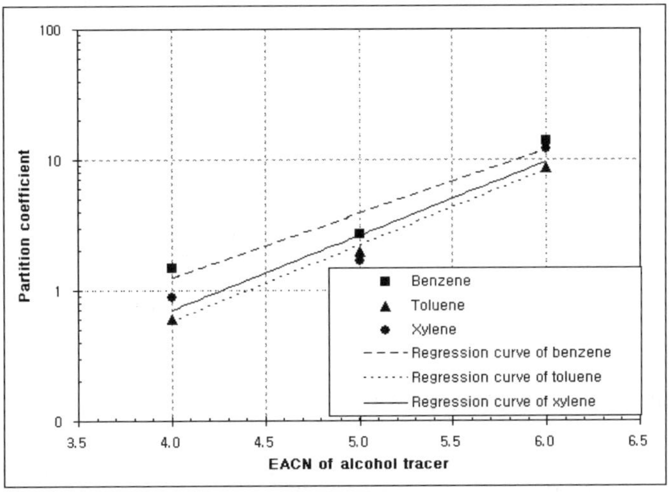

Figure 3: Partitioning coefficient as a function of alcohol EACN.

Table 4: Measured and estimated partition coefficients of alcohol tracers.

Tracers	Mixtures	Exp. logK	Est. logK	Accuracy(%)
Pentanol	Mixture A	0.3729	0.3231	86.65
	Mixture B	0.3962	0.3441	86.86
	Mixture C	0.3636	0.3178	87.41
	Mixture D	0.3747	0.3270	87.26
	Mixture E	0.4594	0.2727	59.36
Hexanol	Mixture A	1.1608	1.0815	93.17
	Mixture B	1.0282	1.0945	106.45
	Mixture C	0.8820	1.0526	119.35
	Mixture D	1.2608	1.0976	87.05
	Mixture E	1.2852	1.0394	80.87
2-Ethyl-1-butanol	Mixture A	0.8432	0.8669	102.80
	Mixture B	0.8287	0.9135	110.24
	Mixture C	0.9069	0.8608	94.92
	Mixture D	0.8169	0.8448	103.42
	Mixture E	0.7589	0.8084	106.53
4-Methyl-2-pentanol	Mixture A	0.6599	0.6083	92.18
	Mixture B	0.6628	0.6609	99.73
	Mixture C	0.7889	0.6003	76.09
	Mixture D	0.6580	0.5770	87.69
	Mixture E	0.9628	0.5548	57.62

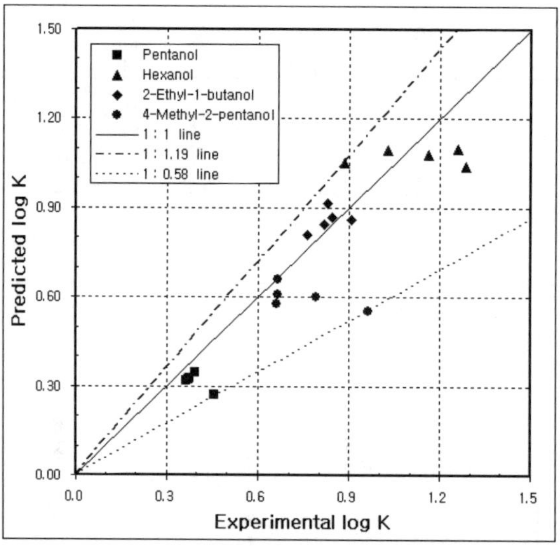

Figure 4: Comparison of measured and estimated partition coefficients of tracers.

4 Summary and conclusion

The linear relation between the concentration of the tracer in water and that of the tracer in BTEX compounds was confirmed with high accuracy (regression coefficients are over 0.95) and partition coefficients of alcohol tracers were obtained under this linear relation.

The bilinear relation was found for correlating the logarithm of the partition coefficient for alcohol tracer partitioning between BTEX compounds and water as a function of the alcohol tracer's EACN and the BTEX EACN. As a result of comparison between experimental and estimated partition coefficients, a general agreement was observed although there is a tendency that the estimated results are underestimated.

Therefore, we can conclude that the EACN approach can be extended to estimate the partition coefficient for alcohol tracer partitioning between BTEX compounds and water.

Acknowledgements

This subject is supported by Korea Ministry of Environment as "The Eco-technopia 21 project".

References

[1] Schade, W. & Bublitz, J. On-site laser probe for the detection of petroleum products in water and soil, *Environ. Sci. Technol.*, **30(5)**, pp. 1451 1458, 1996.

[2] Bachman, J., Kanan, S. M. & Patterson, H. H. Monitoring laboratory-scale bioventing using synchronous scan fluorescence spectroscopy: analysis of vapor phase, *Environmental pollution,* **113**, pp. 155 162, 2001.

[3] Brusseau, M. L., Nelson, N. T., & Cain, R. B. The partitioning tracer method for In-situ detection and quantification of immiscible liquids in the subsurface, *ACS symposium series 725*, 1999

[4] Jin, M., Delshad, M., Dwarakanath, V., McKinney, D. C., Pope, G. A., Sepehrnoori, K., & Tilburg, C. E. Partitioning tracer test for detection, estimation, and remediation performance assessment of subsurface nonaqueous phase liquids, *Water Resources Research,* **31(5)**, pp. 1201 1211, 1995.

[5] Dwarakanath, V. & Pope, G. A. New approach for estimating alcohol partition coefficients between nonaqueous phase liquids and water, *Environ. Sci. Technol.,* **32**, pp. 1662 1666, 1998.

[6] Wu, B. & Sabatini, D. A. Using partitioning alcohol tracers to estimate hydrophobicity of high molecular weight LNAPLs, *Environ. Sci. Technol.,* **34**, pp. 4701 4707, 2000.

[7] Park, H. M., Kim, J. H. Kim, Y. M., & Lee, K. B. BTEX contamination level of soil samples around gas stations in Korea, *Journal of the Korea Society for Environmental Analysis,* **4,** pp. 223 227, 2000.

[8] Cayias, J.L., Schechter, R.S., & Wade, W.H., Modeling Crude Oils for Low Interfacial Tension, *Society of Petroleum Engineers Journal*, **16**, pp. 351 357, 1976.

[9] Baran, J. R., Jr., Pope, G. A., Wade, W. H., Weerasooriya, V., & Yapa, A. Microemulsion formation with mixed chlorinated hydrocarbon liquids, *Journal of Colloid and Interface Science,* **4**, pp. 223 227, 1994.

www.witpress.com, ISSN 1743-3541 (on-line)

Author Index